Grade 8 · Unit 4

Inspire
Science

Humans and Their Place in the Universe

Mc Graw Hill

Phenomenon: What is this cloud made of?

The Carina Nebula is a collection of ionized gases and dust located 7,500 light years from Earth. One of the Milky Way's largest star-forming locations, this nebula stretches over 300 light years.

Fun Fact

The Carina Nebula is home to the superstar binary system Eta Carinae. Eta Carinae is 100 times more massive than the Sun!

FRONT COVER: (t)NASA, ESA, and M. Livio and the Hubble 20th Anniversary Team (STScI), (b)24Novembers/Shutterstock. BACK COVER: NASA, ESA, and M. Livio and the Hubble 20th Anniversary Team (STScI).

mheducation.com/prek-12

Copyright © 2020 McGraw-Hill Education

Send all inquiries to:
McGraw-Hill Education
STEM Learning Solutions Center
8787 Orion Place
Columbus, OH 43240

ISBN: 978-0-07-687532-0
MHID: 0-07-687532-6

Printed in the United States of America.

5 6 7 8 9 10 WEB 25 24 23 22 21 20

McGraw-Hill is committed to providing instructional materials in Science, Technology, Engineering, and Mathematics (STEM) that give all students a solid foundation, one that prepares them for college and careers in the 21st century.

Welcome to

Inspire Science

Explore Our Phenomenal World

Learning begins with curiosity. Inspire Science is designed to spark your interest and empower you to ask more questions, think more critically, and generate innovative ideas.

Start exploring now!

Inspire Curiosity • Inspire Investigation • Inspire Innovation

Authors, Contributors, and Partners

Program Authors

Alton L. Biggs
Biggs Educational Consulting
Commerce, TX

Ralph M. Feather, Jr., PhD
Professor of Educational Studies and
Secondary Education
Bloomsburg University
Bloomsburg, PA

Douglas Fisher, PhD
Professor of Teacher Education
San Diego State University
San Diego, CA

Page Keeley, MEd
Author, Consultant, Inventor of
Page Keeley Science Probes
Maine Mathematics and Science
Alliance
Augusta, ME

Michael Manga, PhD
Professor
University of California, Berkeley
Berkeley, CA

Edward P. Ortleb
Science/Safety Consultant
St. Louis, MO

Dinah Zike, MEd
Author, Consultant, Inventor
of Foldables®
Dinah Zike Academy, Dinah-Might
Adventures, LP
San Antonio, TX

Advisors

Phil Lafontaine
NGSS Education Consultant
Folsom, CA

Donna Markey
NBCT, Vista Unified School District
Vista, CA

Julie Olson
NGSS Consultant
Mitchell Senior High/Second Chance
High School
Mitchell, SD

Content Consultants

Chris Anderson
STEM Coach and Engineering
Consultant
Cinnaminson, NJ

Emily Miller
EL Consultant
Madison, WI

Key Partners

American Museum of Natural History

The American Museum of Natural History is one of the world's preeminent scientific and cultural institutions. Founded in 1869, the Museum has advanced its global mission to discover, interpret, and disseminate information about human cultures, the natural world, and the universe through a wide-ranging program of scientific research, education, and exhibition.

SpongeLab Interactives

SpongeLab Interactives is a learning technology company that inspires learning and engagement by creating gamified environments that encourage students to interact with digital learning experiences. Students participate in inquiry activities and problem-solving to explore a variety of topics through the use of games, interactives, and video while teachers take advantage of formative, summative, or performance-based assessment information that is gathered through the learning management system.

PhET Interactive Simulations

The PhET Interactive Simulations project at the University of Colorado Boulder provides teachers and students with interactive science and math simulations. Based on extensive education research, PhET simulations engage students through an intuitive, game-like environment where students learn through exploration and discovery.

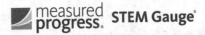

Measured Progress, a not-for-profit organization, is a pioneer in authentic, standards-based assessments. Included with New York Inspire Science is **Measured Progress STEM Gauge**® assessment content which enables teacher to monitor progress toward learning NGSS.

Table of Contents
Humans and Their Place in the Universe

Table of Contents (continued)
Humans and Their Place in the Universe

Earth and Human Activity

ENCOUNTER
THE PHENOMENON

What can patterns of light at night tell us about the human population and resource use?

Earth at Night

GO ONLINE
Watch the video *Earth at Night* to see this phenomenon in action.

Communicate The satellite image shows Earth at night. You can see where the large cities are located. What do you think the dark areas represent? When you turn on the lights at night, where does the energy to power the lights come from? How might this daily activity impact the environment? Record your thoughts in the space below. With a partner, discuss his or her thoughts. Then record what you both would like to share with the class below.

7.6 Billion and Counting

Scientists estimate that more than 7.6 billion people currently live on Earth. It is estimated that by 2050, the human population will reach 9.8 billion. How do the human population and human activities impact Earth?

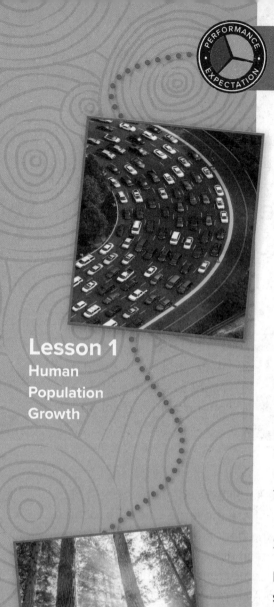

Lesson 1
Human Population Growth

You are on a team of scientists who are trying to answer this question. You will gather evidence to support an argument about how increases in human population and resource use affect Earth's systems. You will present your argument in a panel discussion with your team.

Start Thinking About It

In the image above you see a crowded city street, showing just one tiny portion of the 7.6 billion people on Earth. Think about how many resources you use in one day. How would Earth's systems be affected if one person's daily resource use were multiplied by 7.6 billion people? Discuss your thoughts with your group.

Lesson 2
People and the Environment

STEM Module Project
Planning and Completing the Science Challenge
How will you meet this goal? The concepts you will learn throughout this module will help you plan and complete the Science Challenge. Just follow the prompts at the end of each lesson!

Population Fluctuation

Three friends were arguing about how the human population has changed over time.

Casey: The human population had been fairly constant throughout Earth's history.

Allayna: Population size is increasing at a significant rate due to advances in technology, medicine, and sanitation.

Karl: Population size has been decreasing over time because people are having fewer children.

Which person do you agree with the most? Explain your reasoning.

_____

_____

_____

_____

_____

_____

_____

You will revisit your response to the Science Probe at the end of the lesson.

Human Population Growth

ENCOUNTER
THE PHENOMENON

Why are the numbers of cars on the road increasing, and what's the impact?

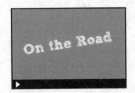

On the Road

GO ONLINE

Check out *On the Road* to see this phenomenon in action.

Using the information in the infographic online, create a graphic organizer that shows the relationships between human population growth and the consumption of natural resources. Display your graphic organizer in the space below.

EXPLAIN
THE PHENOMENON

Are you starting to get some ideas about how increases in the size of the human population causes increases in the consumption of natural resources? Use your observations about the phenomenon to make a claim about how the rate of consumption of natural resources is impacted by human population growth.

CLAIM

The relationship between human population growth and resource consumption is...

 COLLECT EVIDENCE as you work through the lesson. Then return to these pages to record your evidence.

EVIDENCE

A. What evidence have you discovered to explain how the human population has changed over time?

MORE EVIDENCE

B. What evidence have you discovered to explain the impact of population growth on Earth's resources?

When you are finished with the lesson, review your evidence. If necessary, based on the evidence, revise your claim.

REVISED CLAIM

The relationship between human population growth and resource consumption is...

Finally, explain your reasoning for how and why your evidence supports your claim.

REASONING

The evidence I collected supports my claim because...

How fast is the population growing?

You are part of a population of humans. The other species in your area, such as birds or trees, each make up a separate population. A **population** is all the members of a species living in a given area. How has human population changed over time? Let's investigate.

INVESTIGATION

Here We Grow Again

Analyze the information presented in the graph.

Human Population Growth

1. How does the rate of human population growth from years 200 to 1800 compare to the rate of growth from 1800 to 2000?

_____

_____

_____

2. **WRITING ❯ Connection** Research the Industrial Revolution in the 1800s. Then, evaluate the following claim: The Industrial Revolution led to a dramatic population explosion. Construct an argument to support or refute the claim using evidence from your research. Record your response in your Science Notebook.

Human Population When the first American towns were settled, most had low populations. Today, some of those towns are large cities, crowded with people. In a similar way, Earth was once home to relatively few humans. Scientists estimate that there were about 300 million humans on Earth a thousand years ago. Improved health care, clean water, and other technological advancements mean that more people are living longer and reproducing.

As you discovered in the *Here We Grow Again* investigation, the greatest increase in human population occurred during the last few centuries. Today there are more than 7 billion humans on Earth. By 2050 there could be over 9 billion.

COLLECT EVIDENCE

How has human population changed over time? Record your evidence (A) in the chart at the beginning of the lesson.

 GO ONLINE for additional opportunities to explore!

Want to learn more about population growth? Then perform one of the following activities.

☐ **Explore** how the population is projected to change in the **Investigation** *The Human Population.*

OR

☐ Analyze trends in human population data in the **Investigation** *Graphing Population Growth.*

 Want more information?
Go online to read more about how resource availability is affected by a growing population.

FOLDABLES
Go to the Foldables® library to make a Foldable® that will help you take notes while reading this lesson.

Is human population growth equal around the world?

In approximately one hour, 15,000 babies are born worldwide. Technological advances have resulted in a rapid growth in human population. Is human population growth equal in all countries? Let's investigate.

INVESTIGATION

Comparing Ages

Observe the graph showing the percent of youth and elderly in four countries.

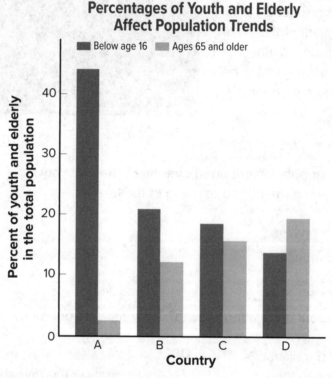

Percentages of Youth and Elderly Affect Population Trends

1. Use the data to predict how the population of each country will change.

_____

_____

_____

_____

_____

_____

2. Brainstorm a list of factors, events, or conditions that affect the growth of human populations in these countries. Predict the effect of each factor on the population growth rate.

_____

_____

_____

_____

Population Structure Populations have a tendency to increase in size. However, many factors influence the rate at which a population can grow. At the simplest level, the rate of increase is determined by subtracting the number of individuals leaving a population from the number entering the population. Individuals leave a population either by death or emigration, which is when people leave an area to settle somewhere else; they enter the population by birth or immigration, which is when people enter an area to live there. Another important characteristic of any population is its age structure. Let's investigate what an age structure diagram can tell us about population growth.

INVESTIGATION

Increasing and Decreasing Populations

The diagram below shows the relative numbers of individuals in pre-reproductive, reproductive, and post-reproductive years for three countries: Kenya, the United States, and Germany.

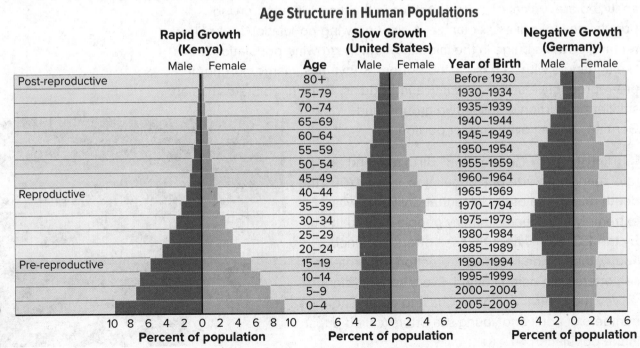

Age Structure in Human Populations

1. How would you describe Kenya's growth?

_____

_____

_____

_____

2. Why does a chart like Germany's show negative growth?

_____

_____

_____

_____

3. What can you determine about population growth rates from observing the top halves of the three charts?

_____

_____

_____

Age Structure A population's age structure is the number of males and females in each of three age groups: pre-reproductive state, reproductive state, and post-reproductive state. The age structure diagrams you just analyzed are typical of many countries in the world. A nongrowing population looks like a rectangle. A slow-growing population looks like a rectangle with a bulge in the middle. A rapidly growing population looks like a triangle with its base at the bottom. Kenya has a large portion of pre-reproductive and reproductive individuals. The United States has a smaller proportion of these two groups, and Germany has an even smaller portion.

LIFE SCIENCE Connection Birth rate and death rate change the size of a population. In the 1700s the death rate of sea otters in central California was extremely high because many people hunted them. By the 1930s only about 50 sea otters remained. Today, the Marine Mammal Protection Act protects sea otters from being hunted. Every spring, scientists survey the central California Coast to determine the numbers of adult and young sea otters (called pups) in the population.

THREE-DIMENSIONAL THINKING

This graph shows the world population growth and projected trends in both developed and developing countries.

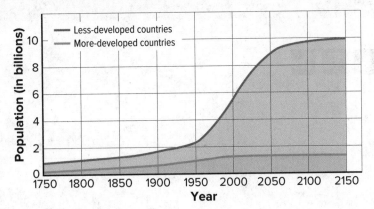

Analyze the data to determine the current population structure of developed and developing countries. Support your reasoning with evidence from the graph.

_____

_____

_____

_____

_____

_____

_____

_____

_____

_____

_____

_____

_____

How does human population growth affect natural resources?

Every human being needs certain things, such as food, clean water, and shelter, to survive. People also need clothes, transportation, and other items. All the items used by people come from resources found on Earth. A **natural resource** is a part of the environment that supplies material useful or necessary for the survival of living things. Let's investigate how natural resources are affected by a growing population.

LAB Bean There, Done That

Safety 🥽 🦺 🧤 🌡️

Materials

dried beans
beaker

Procedure

1. Read and complete a lab safety form.

2. Place 10 dried beans in a 100-mL beaker.

3. At the sounding of each start signal, you will double the number of beans in the beaker. Make a table to record your data. The table should indicate the number of beans added and the total number of beans in the beaker after each addition.

4. Double the number of beans each time the start signal sounds. Take time to record your data between each signal.

5. Follow your teacher's instructions for proper cleanup.

Analyze and Conclude

6. Can you add any more beans to the beaker? Why or why not?

_____

_____

_____

7. How many times did you have to double the beans to fill the beaker?

_____

_____

_____

8. Assess the following statement: The growth of a population does not affect the availability and use of resources, such as space or land. Use evidence from the lab to support your argument.

_____

_____

_____

_____

_____

_____

_____

_____

Population Limits As the human population grows larger and larger, cities become crowded with people, using space and depleting resources. Earth has limited resources. It cannot support a population of any species in a given environment beyond its carrying capacity. **Carrying capacity** is the largest number of individuals of a given species that Earth's resources can support and maintain for a long period of time. Let's explore carrying capacity.

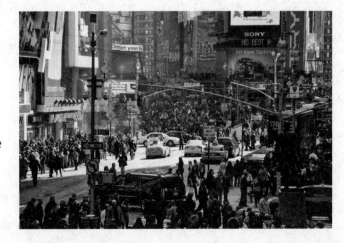

INVESTIGATION

Population Simulation

Kiwi Island is an imaginary 30,000-acre island in the South Pacific. The population of Kiwi Island was stable at 230 people between 2000 and 2010. Many areas of the island were tropical forests uninhabited by people. As more and more people discovered this paradise, the population began to double each year after 2010.

Welcome to Kiwi Island

1. Plot the population growth of Kiwi Island from 2000 to 2017.

2. **MATH Connection** Determine the number of acres available per person for each year that is plotted.

3. **MATH Connection** Describe the relationship between land availability in 2000 and 2017.

_____

_____

4. Predict what happens to resources, such as land, if the population is below its carrying capacity. Explain your reasoning.

_____

_____

5. Predict what happens to resources if the population is reaching or above its carrying capacity. How might this impact the population growth of the island?

_____

_____

Approaching Carrying Capacity Each person uses space and resources. Population size depends on the amount of available resources and how members of the population use them. When population density is low, resources are abundant, and population increases. If resources become scarce or if the environment is damaged, members of the population can suffer and population size can decrease. If the human population continues to grow beyond Earth's carrying capacity, eventually Earth will not have enough resources to support humans. Let's investigate the relationship between resource use and population growth.

INVESTIGATION

Resource Consumption

The graph on the right shows the estimated amount of land required to support a person through his or her lifetime. This includes the land for the production of food, forest products and housing, and the additional forest land required to absorb the carbon dioxide produced by the burning of fossil fuels. A hectare of land is equivalent to approximately 2.5 acres or 10,000 square meters. Take a moment to compare and contrast the estimated hectares of land required to support individuals in each of the listed countries.

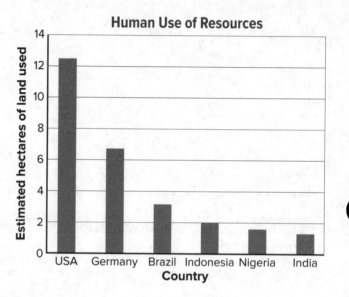

1. In which country is the per-capita consumption the greatest? In which country is the per-capita consumption the least? Why do you think human use of resources varies among countries?

_____

_____

_____

_____

_____

_____

_____

2. MATH ▶Connection Now, calculate the population growth rate (PGR) for each of these countries. Record this information in the table below. Use the following equation.

$$\frac{\text{birth rate} - \text{death rate} + \text{net migration rate}}{10} = \text{PGR (\%)}$$

Population Growth Rates					
Country	Population 2015 (millions)	Birth rate	Death Rate	Net Migration Rate	PGR (%)
*United States	321.2	13	8	3	
*Germany	81.1	8	11	5	
Brazil	204.5	15	6	0	
Indonesia	255.7	21	6	−1	
Nigeria	181.8	39	14	0	
India	1,314.1	21	7	−1	

*Industrially developed country

Source: 2015 Population Reference Bureau

3. How do population growth rates compare in industrially developed countries and developing countries?

_____

_____

4. Compare this data with the Human Use of Resources graph on the previous page. Then evaluate the following statement: Countries with higher population growth rates have a greater impact on resource use than countries with lower population growth rates. Support or refute your argument with evidence.

_____

_____

_____

_____

5. Some developing countries, such as India, are becoming more industrialized and are also increasing in population. Predict the effect that this might have on the carrying capacity of Earth.

_____

_____

_____

_____

_____

_____

Growing Populations and Resource Use Another important factor in keeping the human population at or below the carrying capacity is the amount of resources from the biosphere that are used by each person. Currently, individuals in industrially developed countries use far more resources than those individuals in developing countries. Countries such as India are becoming more industrialized, and they have a relatively high growth rate. These countries are adding more people and are increasing their use of resources.

At some point, the land needed to sustain each person on Earth might exceed the amount of land that is available. Resources become harder to extract and find. However, a decline in these resources often leads to development of new technologies and alternative resources, which in turn expands Earth's carrying capacity. This allows for the population to increase again. But there is a cost to increasing populations. In the following lesson, we will examine how increases in human populations and per-capita consumption of natural resources impact Earth's systems.

THREE-DIMENSIONAL THINKING

Write a paragraph **explaining** why it is important for humans to maintain a population at or below the carrying capacity. Write your response in your Science Notebook.

COLLECT EVIDENCE

What impact does population growth and per-capita consumption have on Earth's resources? Record your evidence (B) in the chart at the beginning of the lesson.

A Day in the Life of a Demographer

Have you ever heard the phrase *population explosion*? Population explosion describes the sudden rise in human population that has happened in recent history. The population has increased at a fairly steady rate for most of human history. In the 1800s, the population began to rise sharply. What kind of scientist studies this information?

A demographer is a scientist who studies populations, particularly human beings. Statistical data such as birth rates, death rates, migration rates, and occurrences of disease help these scientists illustrate the changing size, structure, and distribution of human populations.

A demographer also makes observations about the causes and effects of population changes based on collected data. For example, if an area is experiencing an increase in birth rates or immigration, they will analyze trends and predict future needs for the community.

It's Your Turn

Research Imagine you are a demographer for your city. Research how the population of your city has changed over the past 50 years. Create a short slideshow to predict how the needs of your community have changed based on the data you collect.

Summarize It!

1. **Organize** Complete the Venn diagram below. Below each term write its definition. Where circles overlap, explain the relationship among the terms.

Population

Resource

Carrying Capacity

Three-Dimensional Thinking

Demographers often use age structure diagrams to study human populations in different countries. The shapes of these diagrams provide clues about how quickly the populations are growing. Compare the age distribution of males in the United States and Mexico.

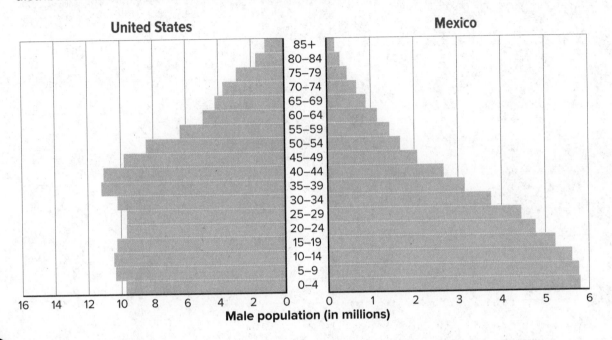

United States / **Mexico**

Male population (in millions)

2. What can be concluded from the data displayed on the age structure diagrams above?

   A The male population of Mexico is growing at a slower rate than the United States because fewer children are being born in Mexico.

   B The male population of Mexico is growing at a faster rate than the United States because the number of children in Mexico exceeds the other age groups.

   C The male population of Mexico is growing at a slower rate than the United States because there are fewer elderly males.

   D The male population of Mexico is growing at a faster rate than the United States because there is a greater number of elderly males.

Real-World Connection

3. Argue A friend makes the following claim: There are enough resources on Earth to support increases in per-capita consumption rates if human population growth rates stay the same. Do you agree? Use evidence and reasoning to support or refute your friend's claim.

_____

_____

_____

_____

_____

_____

_____

_____

 Still have questions?
Go online to check your understanding about human population growth.

REVISIT
SCIENCE PROBES

Do you still agree with the student you chose at the beginning of the lesson? Return to the Science Probe at the beginning of the lesson. Explain why you agree or disagree with that student now.

EXPLAIN
THE PHENOMENON

Revisit your claim about the relationship between resource availability and human population growth. Review the evidence you collected. Explain how your evidence supports your claim.

START PLANNING
STEM Module Project
Science Challenge

Now that you have learned about human population growth and resource availability, go to your Module Project to begin preparing your argument. Keep in mind that you want to argue how the per-capita consumption of natural resources are impacted by human population growth.

Natural Resources

Four friends argued about natural resources and their impact on the environment. This is what they said:

Kate: It is better to use natural resources because they don't harm our environment like human-made resources.

Clint: It is better to use human-made resources because they don't harm our environment like natural resources.

Abby: It doesn't matter—both natural and human-made resources can harm the environment.

Troy: Neither human-made nor natural resources are harmful. They are both good for the environment.

Which friend do you agree with the most? Explain why you agree.

_____

_____

_____

_____

_____

You will revisit your response to the Science Probe at the end of the lesson.

People and the Environment

ENCOUNTER
THE PHENOMENON | How does cutting down trees impact Earth's systems?

▶ GO ONLINE

Watch the video *Deforestation* to see this phenomenon in action.

You have probably seen a tree lying on the forest floor that fell from natural causes. Trees are also cut down for human purposes. It may seem obvious that cutting down trees affects the forests and the other organisms that live there, but did you know each of Earth's systems is impacted when forests are removed? John Muir, an early naturalist for whom Muir Woods is named, once stated: "Tug on anything at all and you'll find it connected to everything else in the universe." With your partner, brainstorm how activities like cutting down forests impacts each of Earth's systems, and how we might mitigate these effects.

EXPLAIN
THE PHENOMENON

Are you starting to get some ideas about how removing forests impacts Earth? Use your observations about the phenomenon to make a claim about how the consumption of resources impacts Earth's systems and how we might mitigate these impacts.

CLAIM

The consumption of natural resources, like trees, impacts... and can be mitigated by...

COLLECT EVIDENCE as you work through the lesson.
Then return to these pages to record your evidence.

EVIDENCE

A. What evidence have you discovered to explain why humans remove entire forests?

B. What evidence have you discovered to explain how the removal of trees impacts Earth's systems?

MORE EVIDENCE

C. What evidence have you discovered to explain how we can mitigate these impacts?

When you are finished with the lesson, review your evidence. If necessary, based on the evidence, revise your claim.

REVISED CLAIM

The consumption of natural resources, like trees, impacts... and can be mitigated by...

Finally, explain your reasoning for how and why your evidence supports your claim.

REASONING

The evidence I collected supports my claim because...

Why cut down trees?

Many experts are concerned about the loss of forest cover worldwide. The removal of large areas of forests for human purposes is called **deforestation.** Why do humans cut down trees?

Falling Forests

1. **MATH Connection** In 1630 at the beginning of European settlement, about 425,000,000 hectares of the United States were forested. By 2017, the area of forested land had declined to 300,000,000 hectares. What is the percent decrease of forested land between 1630 and 2017?

2. **MATH Connection** The population of the United States in 1630 was around 4,500. In 2017, the population was around 326,500,000. What percent increase does this represent?

3. What cause-and-effect relationship can you identify between the removal of forests and human population growth?

_____

_____

DrimaFilm/Shutterstock.com

Copyright © McGraw-Hill Education

Want more information?
Go online to read more about how human populations impact Earth's systems.

FOLDABLES
Go to the Foldables® library to make a Foldable® that will help you take notes while reading this lesson.

Urbanization As you have discovered, increases in the size of the human population cause increases in the consumption of natural resources. Trees are cut down for fuel, paper products, and wood products. People also clear forests for urban development. **Urbanization**—the development of towns and cities on previously natural areas—is the most human-dominated and fastest-growing type of land use worldwide.

Agriculture Exploding population growth on our planet and the amount of land available to produce food for that population are not keeping pace with each other. Forest clearing for agriculture helps answer the question of how the world's growing human population can be fed. Brazil's Cerrado represents one of the world's last opportunities to open a large area of new, highly productive cropland. The Cerrado, 2 million hectares of grassland and tropical forest, is the site of the world's fastest-growing soybean production.

Rapid growth of soy production in Brazil has both positive and negative aspects. On one hand, more food is now available to feed the world. On the other hand, the rapid expansion of agriculture worldwide—at the expense of forests—has environmental effects on Earth's systems.

COLLECT EVIDENCE
Why do people remove entire forests? Record your evidence (A) in the chart at the beginning of the lesson.

How does deforestation impact Earth's systems?

Nearly half of the world's tropical deforestation occurs in the "arc of deforestation" between the Cerrado and the Amazon rain forest. Let's take a closer look.

Arc of Deforestation

Study the map below. Then answer the questions that follow.

1. The continent's highest rate of forest clearing is in the "arc of deforestation." Why do you think these forests are cleared? Predict how this might affect the geosphere.

_____

_____

_____

2. **LIFE SCIENCE** ▶ **Connection** Biologically, the Cerrado is the richest savanna in the world, with at least 130,000 plant and animal species. The Amazon rain forest is estimated to host some 10 percent of the planet's species. Predict the effects that agricultural expansion might have on the plant and animal species that live there.

_____

_____

_____

3. Where on the map does deforestation strongly impact the hydrosphere? Explain.

_____

_____

4. Finally, predict how a reduction in trees might impact the atmosphere. Record your thoughts below.

_____

_____

_____

Changes in Earth's Systems The consumption of natural resources, like trees and land, has environmental effects of global importance. Forest clearing for cropland and pasture diminishes habitat. Habitat loss is often the greatest extinction threat for the biosphere.

Clearing forests can also affect the geosphere and hydrosphere. Plant roots hold soil in place. Without these natural anchors, soil erodes away. Drought can also follow deforestation. Cleared land dries rapidly and stores little moisture.

And finally, deforestation also impacts the atmosphere. Trees remove carbon dioxide from the atmosphere during photosynthesis. Rates of photosynthesis decrease when large areas of trees are cut down, and more carbon dioxide remains in the atmosphere.

THREE-DIMENSIONAL THINKING

Outline an **argument** for how deforestation impacts Earth's **systems**. What are the short-term consequences of clearing land? Using your knowledge of the **cause-and-effect** relationships associated with deforestation, predict the long-term consequences of clearing entire forests.

_____

_____

_____

_____

_____

_____

_____

COLLECT EVIDENCE

How does deforestation impact Earth's systems? Record your evidence (B) in the chart at the beginning of the lesson.

Deforestation and Carbon Dioxide in the Atmosphere

How does carbon dioxide affect climate?

What do you think when you hear the words *greenhouse gases*? Many people picture pollution from automobiles or factory smokestacks from the consumption of energy resources. It might be surprising to learn that cutting down forests affects the amount of one of the greenhouse gases in the atmosphere—carbon dioxide.

Trees, like most plants, carry out photosynthesis and make their own food. Carbon dioxide from the atmosphere is one of the raw materials of photosynthesis. When deforestation occurs, there are fewer trees to remove carbon dioxide from the atmosphere. As a result, the level of carbon dioxide in the atmosphere increases.

Trees affect the amount of atmospheric carbon dioxide in other ways. Large amounts of carbon are stored in the molecules that make up trees. When trees are burned or left to rot, much of this stored carbon is released as carbon dioxide. This increases the amount of carbon dioxide in the atmosphere.

Carbon dioxide in the atmosphere has an impact on climate. Greenhouse gases, such as carbon dioxide, increase the amount of the Sun's energy that is absorbed by the atmosphere. They also reduce the ability of thermal energy to escape back into space. So, when levels of carbon dioxide in the atmosphere increase, more thermal energy is trapped in Earth's atmosphere. This can lead to climate change.

▲ These cattle are grazing on land that was once part of a forest in Brazil.

▲ In a process called slash-and-burn, forest trees are cut down and burned to clear land for agriculture.

It's Your Turn

WRITING Connection What are the short-term and long-term consequences of climate change? Draw evidence from research to support your analysis. Create a video blog in which you cite specific textual evidence to reinforce your argument.

What can be done?

Can Earth sustain our current lifestyles? Will there be adequate natural resources for future generations? These questions are among the most important in environmental science today. We depend on nature for food, water, energy, oxygen, and other systems that support life. For resource use to be sustainable, we cannot consume resources faster than nature can replenish them. Degrading ecological systems ultimately threatens everyone's well-being. Therefore, scientists, governments, and concerned citizens around the world are working to identify environmental problems, educate the public about them, and help find solutions.

⚙ENGINEERING INVESTIGATION

Engineering Solutions

Technologies, such as the conservation drone pictured here, can help monitor and minimize human impact on the environment. Research two technologies used to reduce the effects of human populations on Earth's systems. Describe how these engineered solutions work in the space provided.

Copyright © McGraw-Hill Education   Dmitry Kalinovsky/
Shutterstock.com

Call to Action Threats to supplies of food, energy, water, and nature certainly require technologies and action on national and international levels. However, in each of these cases, ethically responsible action can also begin with the individual.

Personal Choices The concept of an ecological footprint has been developed to help individuals measure their environmental impact on Earth. Your **ecological footprint** is defined as the area of Earth's productive land and water required to supply the resources that an individual demands as well as to absorb the wastes that the individual produces. You can calculate your own footprint by answering a simple questionnaire about consumption patterns, such as electricity use, shopping, and driving habits.

INVESTIGATION

Step Toward a Better Future

1. Calculate your ecological footprint score using the quiz provided by your teacher.

2. How many planet Earths does it take to support your lifestyle?

3. In the space below, brainstorm actions you could take to minimize your ecological footprint.

4. **HISTORY** **Connection** Think about the words of anthropologist Margaret Mead, who stated, "Never doubt that a small group of thoughtful, committed citizens can change the world. Indeed, it's the only thing that ever has." What do you think Mead was trying to draw attention to?

_____

_____

THREE-DIMENSIONAL THINKING

How can the ecological footprint **model** be used to predict how changes in resource consumption might **effect** Earth's **systems?**

_____

_____

_____

_____

_____

_____

Sustainable Resource Use Technological change sometimes can reduce our ecological footprint. For example, world food production has increased about fourfold since 1950, mainly through advances in irrigation, fertilizer use, and higher-yielding crop varieties, rather than through increased croplands. Similarly, switching to renewable energy sources such as wind and solar power can greatly reduce our ecological footprint. In Germany, which has invested heavily in wind, solar, small-scale hydropower, and public transportation, the ecological footprint is only 4.6 global hectares (gha) per person. In comparison, the average resident of the United States lives at a consumption level that requires 7.2 gha of biologically productive land.

COLLECT EVIDENCE

What types of technologies and activities help mitigate the effects of changes in Earth's systems? Record your evidence (C) in the chart at the beginning of the lesson.

Read a Scientific Text

You have learned that as the population increases, so does our impact on the planet. However, while science can describe the consequences of actions, science cannot necessarily prescribe the decisions that society takes.

Inspect

Read the passage *Should Environmental Scientists Be Advocates for Environmental Policy?*

Find Evidence

Reread the second paragraph. Highlight the arguments for and against having environmental scientists as advocates for environmental policy.

Make Connections

Communicate What are your thoughts? With your partner, discuss whether you think environmental scientists should be policy advocates. Does involvement in policy issues compromise the role of a scientist or enhance it?

Should Environmental Scientists Be Advocates for Environmental Policy?

Should environmental scientists be advocates for environmental policy? This question has been at the heart of policy debates since the environment became a major policy agenda item. A growing number of environmental scientists are now answering "yes." They argue that scientists, by virtue of being citizens first and scientists second, have a responsibility to advocate to the best of their abilities and in a justified and transparent manner.

Much of what has been written about advocacy looks at its appropriateness without adequately assessing its nature. For example, most of the arguments, whether they are for or against advocacy, often look at only one side of the debate. The question of scientific credibility is often raised. Those who oppose scientists acting as advocates often say that advocacy undermines a scientist's credibility. The counterargument could say that as long as a scientist's work is transparently honest, the scientific community is obligated to support it.

There is one area about which many scientists do agree: that as citizens first and scientists second, scientists have a responsibility to use their scientific data and insights to guide policy decisions. A growing number of scientists are calling for more active participation of their profession in matters of policy. They argue that broad participation of scientists in policy issues will undoubtedly result in disagreement among good scientists and will complicate the policy-making process. It is further argued, however, that the ultimate goal should not be simplicity in the process but rather the betterment of society. Many scientists state that the risk of not participating in policy debates on environmental issues is greater than the risk of participating.

Review

Summarize It!

1. **Sketch** Create a concept sketch that describes your understanding of how changes in human populations have a causal role in changing Earth's systems. To construct a concept sketch, being by listing the relationships you want to describe. Then, draw your sketch and write complete sentences describing the sketch. Be creative!

Three-Dimensional Thinking

Biologists report that habitat destruction, overexploitation, and pollution are eliminating species at a rate comparable to the great extinction that marked the end of the age of dinosaurs. The UN Environment Programme reports that, over the past century, more than 800 species have disappeared and at least 10,000 species are now considered threatened. This includes about half of all primates and around 10 percent of all plant species. At least half of the forests existing before the introduction of agriculture have been cleared, and much of the diverse "old growth" on which many species depend for habitat is rapidly being cut and replaced by ecologically impoverished forest plantations.

2. Which statement is a logical conclusion that can be drawn from information in the text?

   A Clean technology has helped eliminate pollution and protect endangered species.

   B Natural resources are limited, and the biosphere has limited capacity to provide our food and water.

   C Consuming forest resources interferes with the survival of other species.

   D "Old-growth" forests supply food, energy, and other resources humans depend on.

---

3. **MATH Connection** Only a small percentage of Americans owned cars before the 1940s. By 2017, there were nearly 250 million vehicles for 323 million people, greatly increasing the need for roadways. In 1960, the United States had about 16,000 km of interstate highways. Today, the interstate highway system includes 77,000 km of paved roadways. What percent increase does this represent?

   A 381 percent

   B 792 percent

   C 38 percent

   D 79 percent

Real-World Connection

4. **Evaluate** the impact on land, water, and/or the atmosphere of an activity you perform daily that requires the consumption of natural resources. How could you minimize these effects?

_____

_____

_____

_____

_____

_____

_____

_____

_____

 Still have questions?
Go online to check your understanding about how human populations impact Earth's systems.

 REVISIT
PAGE KEELEY
SCIENCE PROBES

Do you still agree with the friend you chose at the beginning of the lesson? Return to the Science Probe at the beginning of the lesson. Explain why you agree or disagree with that friend now.

PLAN AND PRESENT
STEM Module Project
Science Challenge

Now that you've learned about how humans impact the environment, go back to your Module Project to continue planning your argument to present in a panel discussion with your team. Your goal is to explain how increases in the human population and consumption of natural resources impact Earth's systems.

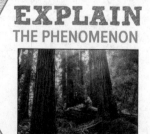 **EXPLAIN**
THE PHENOMENON

Revisit your claim about how natural resource consumption causes changes in Earth's systems. Review the evidence you collected. Explain how your evidence supports your claim.

7.6 Billion and Counting

Scientists estimate that more than 7.6 billion people currently live on Earth. By 2050, the human population is projected to reach 9.8 billion. How do the human population and human activities impact Earth?

You are on a team of scientists who are trying to answer this question. You will gather evidence to support an argument about how increases in human population and resource use affect Earth's systems. You will present your argument in a panel discussion with your team.

Planning After Lesson 1

In the space below, list natural resources that you use every day. As a team, select a natural resource to investigate. Explain why you selected that particular natural resource.

Planning After Lesson 1, continued

Next, select a particular region to research. In the space below, construct a graph that shows changes in the human population of the region over a given time.

Conduct research on the per-capita consumption of your resource by the human population in your chosen region. If possible, obtain information about past, current, and future rates of usage. Organize your findings in a chart or graph.

Compare your population data and your resource consumption data. Recognize and represent any patterns or relationships you see in your Science Notebook.

Planning After Lesson 2

How is your natural resource obtained or processed? How is it used?

How does the use of your resource change the appearance, composition, and/or structure of Earth's systems in your region? Include both direct and indirect impacts.

Research some solutions engineered to alter the impacts of the use of your resource on Earth systems. Describe the solutions in your Science Notebook.

Construct Your Argument

Construct an argument supported by evidence and scientific reasoning for how increases in human population and per-capita consumption of natural resources impact Earth's systems. Include information about engineered solutions for these impacts.

Evaluate and Critique Your Evidence

After you write your scientific argument, take time to evaluate and critique the evidence you used to support your argument.

	Yes	No	Proposed Solution
Is your evidence necessary and sufficient for supporting your claim?			
Is your evidence sufficient to determine causal relationships between consumption of the resource and the impact on Earth's systems?			
Did you consider alternative interpretations of the evidence?			
Can you describe why the evidence supports your claim, as opposed to any alternative claims?			

Present Your Argument

After you have written and evaluated your argument, answer the questions below. Then present your argument in a panel discussion with your team.

In the space below, summarize the relationship among the growth of human populations, resource use, and impacts on Earth's systems.

Review the solutions proposed to alter the effects of resource use on Earth's systems. Propose an improved solution to better reduce any negative effects on one or more Earth's systems.

Use what you've learned to evaluate this statement: Science can determine which decisions society makes about resource use.

Congratulations!
You've completed the
Science Challenge
requirements!

Module Wrap-Up

REVISIT
THE PHENOMENON

Using the concepts you've learned in the module, explain how patterns of light at night can reveal relationships among human population growth, resource use, and the resulting environmental impacts.

OPEN INQUIRY

What are one or two questions you still have about the phenomenon?

_____

_____

_____

_____

Choose the question you are most interested in. Plan and conduct an investigation to answer this question.

The Sun-Earth-Moon System

How do the movements of the Sun, Earth, and the Moon influence the seasons, phases of the Moon, and eclipses?

Sunrise Over Earth

GO ONLINE
Watch the video *Sunrise Over Earth* to see this phenomenon in action.

Communicate Think about seasons, eclipses, and the phases of the Moon. Record your ideas for how or why you think these events happen below. Discuss your ideas with three different partners. Revise or update your ideas, if necessary, after the discussions with your classmates.

Patterns in the Sky

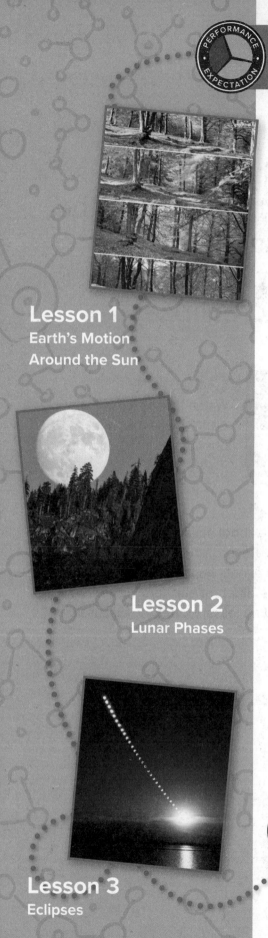

Imagine you are an astronomer who works for a science museum. A group of third-grade students is visiting the museum to learn more about the Sun-Earth-Moon system.

You have been asked to develop a model to help students visualize the patterns of lunar phases, eclipses of the Sun and the Moon, and seasons. To test your model, you will use it to predict different patterns in the sky.

Lesson 1
Earth's Motion Around the Sun

Lesson 2
Lunar Phases

Start Thinking About It

In the image above you see an eclipse of the Sun. How do scientists use models when studying eclipses, lunar phases, and seasons? Why do scientists use models?

Lesson 3
Eclipses

STEM Module Project
Planning and Completing the Science Challenge
How will you meet this goal? The concepts you will learn throughout this module will help you plan and complete the Science Challenge. Just follow the prompts at the end of each lesson!

Seasons

Six friends wondered why it is hot in the summer and cold in the winter. This is what they said:

Mario: I think Earth is closer to the Sun in the summertime.

Louie: I think the Sun gives off more heat in the summertime.

Brad: I think Earth rotates toward the Sun in summertime.

Chelsea: I think Earth tilts towards the Sun in summer and away from the Sun in winter.

April: I think the sunlight that hits Earth is spread out more in the summertime.

Pedro: I don't think it is any of things you described. I think it is something else.

Which friend do you agree with the most? _____ Explain why you agree with that friend.

_____

_____

_____

_____

You will revisit your response to the Science Probe at the end of the lesson.

Earth's Motion Around the Sun

●ENCOUNTER
THE PHENOMENON | What causes the seasons?

Imagine you and your family go hiking in the same woods every year during each season. At noon you always stop for a picnic lunch at the same clearing at the top of the same hill. You always noticed that the scenery changes from spring to summer to fall to winter, but one year you notice that the Sun's apparent path across the sky changes from season to season. The Sun appears lowest in the sky during the winter and highest in the sky during the summer. Diagram why you think the Sun's position in the sky changes each season in the space below. Then, support your diagram with your reasoning.

Seasons

▶ GO ONLINE
Watch the video *Seasons* to see this phenomenon in action.

ENCOUNTER
THE PHENOMENON

In the previous activity, you diagrammed why the Sun appears to change location in the sky over the course of a year. How might this phenomena relate to changing seasons? Use your observations about the phenomenon to make a claim about what causes seasons.

CLAIM

Seasons are caused by...

 COLLECT EVIDENCE as you work through the lesson. Then return to these pages to record your evidence.

EVIDENCE

A. What evidence have you discovered to explain how Earth moves?

B. What evidence have you discovered to explain how the curve of Earth's surface affects temperatures on Earth?

MORE EVIDENCE

C. What evidence have you discovered to explain the relationship between Earth's tilt and seasons?

When you are finished with the lesson, review your evidence. If necessary, based on the evidence, revise your claim.

REVISED CLAIM

Seasons are caused by...

Finally, explain your reasoning for how and why your evidence supports your claim.

REASONING

The evidence I have collected supports my claim because...

How does Earth move?

If you look outside at the ground, trees, and buildings, it does not seem like Earth is moving. Yet Earth is always in motion, spinning in space and traveling around the Sun. Earth's motion changes how energy from the Sun spreads out over Earth's surface. Let's investigate further.

INVESTIGATION

Night and Day

1. Have you ever thought about the causes of night and day? Observe the demonstration your teacher performs and record your observations, including drawings of what you observe.

2. Do you think every planet in the solar system has night and day? Why or why not?

_____

_____

_____

_____

Want more information?
Go online to read more about Earth's motion around the Sun.

FOLDABLES
Go to the Foldables® library to make a Foldable® that will help you take notes while reading this lesson.

Earth's Rotation As Earth moves around the Sun, it spins. A spinning motion is called **rotation.** Some spinning objects rotate on a rod or axle. Earth rotates on an imaginary line through its center. The line on which an object rotates is the **rotation axis.**

Suppose you could look down on Earth's North Pole and watch Earth rotate. You would see that Earth rotates on its rotation axis in a counterclockwise direction, from west to east. One complete rotation of Earth takes about 24 hours. This rotation helps produce Earth's cycle of day and night.

Earth's Revolution Earth moves around the Sun in nearly a circular path. The path an object follows as it moves around another object is an **orbit.** The motion of one object around another is called **revolution.** Earth makes one complete revolution around the Sun every 365.24 days.

Why does Earth orbit the Sun? The answer is the Law of Universal Gravitation. This law states that the pull of gravity between two objects depends on the masses of the objects and the distance between them. The more mass either object has, or the closer together they are, the stronger the gravitational pull.

Earth's Tilted Axis Earth's rotation axis is tilted. The tilt of Earth's rotation axis is always in the same direction by the same amount. This means that during half of Earth's orbit, the north end of the rotation axis is tilted toward the Sun. During the other half of Earth's orbit, the north end of the rotation axis is tilted away from the Sun.

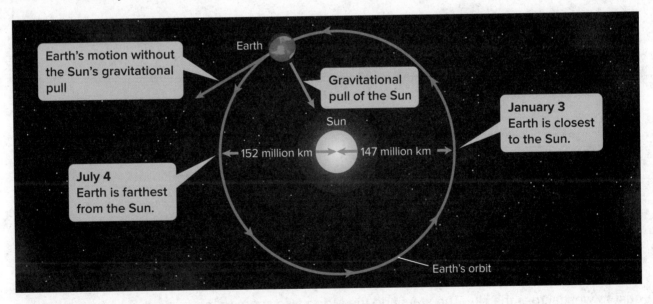

Earth

Earth's motion without the Sun's gravitational pull

Gravitational pull of the Sun

Sun

January 3 Earth is closest to the Sun.

152 million km   147 million km

July 4 Earth is farthest from the Sun.

Earth's orbit

The Sun's effect on Earth's motion is illustrated above. Earth's motion around the Sun is like the motion of an object twirled on a string. The string pulls on the object and makes it move in a circle. If the string breaks, the object flies off in a straight line. In the same way, the pull of the Sun's gravity keeps Earth revolving around the Sun in a nearly circular orbit. If the gravity between Earth and the Sun were to somehow stop, Earth would fly off into space in a straight line.

Why does the view of the sky change over time?

Have you ever looked up at the sky on a clear, dark night and seen stars? If you look at a clear night sky for a long time, what do you notice about the stars?

INVESTIGATION

Star Gazing

⬢ GO ONLINE Watch the video *Star Gazing*.

Record your observations about the stars. Describe what you think is happening. Include drawings and illustrations to help explain your observations.

Apparent Motion Each day the Sun appears to move from east to west across the sky. It seems as if the Sun is moving around Earth. However, it is Earth's rotation that causes the Sun's apparent motion.

Earth rotates from west to east. As a result, the Sun appears to move from east to west across the sky. The stars and the Moon also seem to move from east to west across the sky due to Earth's west to east rotation.

To better understand this, imagine riding on a merry-go-round. As you and the ride move, people on the ground appear to be moving in the opposite direction. In the same way, as Earth rotates from west to east, the Sun appears to move from east to west.

What evidence have you discovered to explain how Earth moves?
Record your evidence (A) in the chart at the beginning of the lesson.

Why is Earth warmer at the equator and colder at the poles?

Temperatures near Earth's poles are colder than temperatures near the equator. What causes these temperature differences? Let's find out!

INVESTIGATION

Ahead of the Curve

Observe the demonstration your teacher conducts using a globe and a flashlight. Illustrate and describe what you observe below.

1. Compare and contrast the shapes you drew above.

_____

_____

_____

_____

2. At which location on the globe is the light more spread out? Explain your response.

_____

_____

_____

_____

Temperature and Earth's Curved Surface As Earth orbits the Sun, only one half of Earth faces the Sun at a time. A beam of sunlight carries energy. The more sunlight that reaches a part of Earth's surface, the warmer that part becomes. Because Earth's surface is curved, different parts of Earth's surface receive different amounts of the Sun's energy.

Energy Received by a Tilted Surface Suppose you shine a beam of light on a flat card. As you tilt the card relative to the direction of the light beam, light becomes more spread out on the card's surface. As a result, the energy that the light beam carries also spreads out more over the card's surface. An area on the surface within the light beam receives less energy when the surface is more tilted relative to the light beam.

The Tilt of Earth's Curved Surface Instead of being flat like a card, Earth's surface is curved. Relative to the direction of a beam of sunlight, Earth's surface becomes more tilted as you move north or south from the equator. As shown below, the energy in a beam of sunlight tends to become more spread out the farther you travel north or south from the equator. This means that regions near the poles receive less energy than regions near the equator. This makes Earth colder at the poles and warmer at the equator.

Sunlight is more spread out away from equator.

Sunlight is less spread out near equator.

Equator

Sunlight is more spread out away from equator.

Based on your observations from the *Ahead of the Curve* investigation and the **model** you used, how is the Sun's energy received on Earth? What **effect** does Earth's curved surface have on temperatures?

_____

_____

_____

_____

_____

_____

_____

COLLECT EVIDENCE

What evidence have you discovered to explain how the curve of Earth's surface affects temperatures on Earth? Record your evidence (B) in the chart at the beginning of the lesson.

Why do Earth's seasons change as Earth orbits the Sun?

You might think that summer occurs in the northern hemisphere when Earth is closest to the Sun, and winter occurs in the southern hemisphere when Earth is farthest from the Sun. However, seasonal changes do not depend on Earth's distance from the Sun. In fact, Earth is closest to the Sun in January! Instead, it is the tilt of Earth's rotation axis, combined with Earth's motion around the Sun, that causes the seasons to change.

Fall and Winter in the Northern Hemisphere
During one half of Earth's orbit, the north end of the rotation axis is away from the Sun. Then, the northern hemisphere receives less solar energy than the southern hemisphere. Temperatures decrease in the northern hemisphere and increase in the southern hemisphere. This is when fall and winter occur in the northern hemisphere, and spring and summer occur in the southern hemisphere.

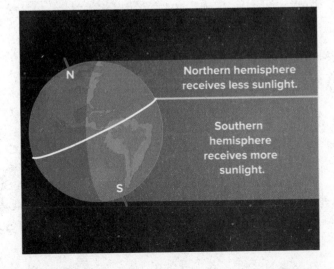

Northern hemisphere receives less sunlight.

Southern hemisphere receives more sunlight.

Spring and Summer in the Northern Hemisphere During the other half of Earth's orbit, the north end of the rotation axis is toward the Sun. Then, the northern hemisphere receives more energy from the Sun than the southern hemisphere. Temperatures increase in the northern hemisphere and decrease in the southern hemisphere. Days last longer in the northern hemisphere, and nights last longer in the southern hemisphere. This is when spring and summer occur in the northern hemisphere, and fall and winter occur in the southern hemisphere.

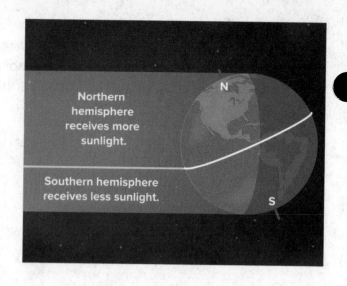

Northern hemisphere receives more sunlight.

Southern hemisphere receives less sunlight.

THREE-DIMENSIONAL THINKING

How does the varying amount of the Sun's solar energy **cause** the seasons? What **effects** does the tilt of Earth's rotation axis have on the seasons?

What is Earth's seasonal cycle?

You've learned that as Earth travels around the Sun, its rotational axis always points in the same direction in space. However, the amount that Earth's rotation axis is toward or away from the Sun changes. This causes changes in a yearly cyclical pattern.

There are four days each year when the direction of Earth's rotation axis is special relative to the Sun. A **solstice** is a day when Earth's rotation axis is the most toward or away from the Sun. An **equinox** is a day when Earth's rotation axis is leaning along Earth's orbit, neither toward nor away from the Sun.

March Equinox to June Solstice When the north end of the rotation axis gradually points more and more toward the Sun, the northern hemisphere gradually receives more solar energy. This is spring in the northern hemisphere.

June Solstice to September Equinox The north end of the rotation axis continues to point toward the Sun but does so less and less. The northern hemisphere starts to receive less solar energy. This is summer in the northern hemisphere.

September Equinox to December Solstice The north end of the rotation axis now points more and more away from the Sun. The northern hemisphere receives less and less solar energy. This is fall in the northern hemisphere.

December Solstice to March Equinox The north end of the rotation axis continues to point away from the Sun but does so less and less. The northern hemisphere starts to receive more solar energy. This is winter in the northern hemisphere.

Changes in the Sun's Apparent Path Across the Sky The figure below shows how the Sun's apparent path through the sky changes from season to season in the northern hemisphere. The Sun's apparent path through the sky in the northern hemisphere is lowest on the December solstice and highest on the June solstice.

The seasons change as Earth moves around the Sun. Earth's motion around the Sun causes Earth's tilted rotation axis to be leaning toward the Sun and away from the Sun. Let's examine what this looks like.

December Solstice

The December solstice is on December 21 or 22.
On this day

- the north end of Earth's rotation axis is away from the Sun;
- days in the northern hemisphere are shortest and nights are longest; winter begins;
- days in the southern hemisphere are longest and nights are shortest; summer begins.

September Equinox

The September equinox is on September 22 or 23.
On this day

- the north end of Earth's rotation axis leans along Earth's orbit;
- there are about 12 hours of daylight and 12 hours of darkness everywhere on Earth;
- autumn begins in the northern hemisphere;
- spring begins in the southern hemisphere.

March Equinox

The March equinox is on March 20 or 21.
On this day

- the north end of Earth's rotation axis leans along Earth's orbit;
- there are about 12 hours of daylight and 12 hours of darkness everywhere on Earth;
- spring begins in the northern hemisphere;
- autumn begins in the southern hemisphere.

June Solstice

The June solstice is on June 20 or 21.
On this day

- the north end of Earth's rotation axis is toward the Sun;
- days in the northern hemisphere are longest and nights are shortest; summer begins;
- days in the southern hemisphere are shortest and nights are longest; winter begins.

Read a Scientific Text

Astronomical seasons, which are based on the position of Earth in relation to the Sun, differ from the meteorological seasons. Read on to find out why!

CLOSE READING

Inspect

Read the passage *Meteorological Versus Astronomical Seasons.*

Find Evidence

Reread the fourth paragraph. Underline the relationships among events that can be described as cause-and-effect relationships.

Make Connections

Communicate With your partner, discuss how meteorological observing led to the creation of meteorological seasons. How do meteorological seasons differ from astronomical seasons?

PRIMARY SOURCE

Meteorological Versus Astronomical Seasons

You may have noticed that meteorologists and climatologists define seasons differently from "regular" or astronomical spring, summer, fall, and winter. So, why do meteorological and astronomical season begin and end at different times? In short, it's because the astronomical seasons are based on the position of Earth in relation to the sun, whereas the meteorological seasons are based on the annual temperature cycle.

People have used observable periodic natural phenomena to mark time for thousands of years. The natural rotation of Earth around the sun forms the basis for the astronomical calendar, in which we define seasons with two solstices and two equinoxes. Earth's tilt and the sun's alignment over the equator determine both the solstices and equinoxes.

The equinoxes mark the times when the sun passes directly above the equator. In the Northern Hemisphere, the summer solstice falls on or around June 21, the winter solstice on or around December 22, the vernal or spring equinox on or around March 21, and the autumnal equinox on or around September 22. These seasons are reversed but begin on the same dates in the Southern Hemisphere.

Because Earth actually travels around the sun in 365.24 days, an extra day is needed every fourth year, creating what we know as Leap Year. This also causes the exact date of the solstices and equinoxes to vary. Additionally, the elliptical shape of Earth's orbit around the sun causes the lengths of the astronomical seasons to vary between 89 and 93 days. These variations in season length and season start would make it very difficult to consistently compare climatological statistics for a particular season from one year to the next. Thus, the meteorological seasons were born.

Meteorologists and climatologists break the seasons down into groupings of three months based on the annual temperature cycle as well as our calendar. ...

Meteorological observing and forecasting led to the creation of these seasons, and they are more closely tied to our monthly civil calendar than the astronomical seasons are. The length of the meteorological seasons is also more consistent, ranging from 90 days for winter of a non-leap year to 92 days for spring and summer. By following the civil calendar and having less variation in season length and season start, it becomes much easier to calculate seasonal statistics from the monthly statistics, both of which are very useful for agriculture, commerce, and a variety of other purposes.

Source: National Oceanic and Atmospheric Administration

THREE-DIMENSIONAL THINKING

When the northern hemisphere is experiencing summer, Earth is the farthest away from the Sun in its orbit. That distance, about 152,000,000 km, is called the aphelion. Approximately how far is Earth from the Sun in miles? **Solve** the conversion **equation** to calculate the distance in miles. Multiply the distance in km by the conversion factor of 0.62 km. Next find the distance in miles of the perihelion—when Earth is closest to the Sun in its orbit. This is when the northern hemisphere experiences winter.

Aphelion in km: about 152,000,000

Aphelion in miles:

Perihelion in km: about 147,000,000

Perihelion in miles:

If the shape of Earth's orbit was unaltered, but the rotational axis was **changed** so that Earth had no tilt with respect to the orbit, predict the **effect** on the seasons.

_____

_____

_____

_____

_____

_____

COLLECT EVIDENCE

How does the tilt of Earth's rotation axis affect the seasons? Record your evidence (C) in the chart at the beginning of the lesson.

A Closer Look: Polar Night & Midnight Sun

Is it possible to have 24 hours of nighttime or 24 hours of sunlight? Parts of Earth experience these natural phenomena called polar night and midnight sun. How does this happen?

Earth completes one rotation on its tilted axis every 24 hours causing day and night. Because of Earth's tilt, the Sun may never set or may never rise at certain locations on Earth. If the Arctic Circle is tilted toward the Sun, then areas in the Arctic Circle will experience six months of daylight, where the Sun does not set. This begins when the vernal equinox occurs.

While the Arctic Circle experiences midnight sun, the Antarctic Circle will experience polar night. During polar night, the Sun does not rise for six months because the Antarctic Circle is tilted away from the Sun.

Once the autumnal equinox occurs, the opposite phenomenon occurs in each location. The Arctic Circle will experience polar night while the Antarctic Circle will experience midnight sun.

In the northern hemisphere, polar night and midnight sun occur in Norway, parts of Alaska, Canada, Greenland, Finland, Russia, and Sweden. The only place this occurs in the southern hemisphere is in Antarctica.

It's Your Turn

HEALTH ▶ Connection   How does living six months with 24 hours of sunlight or darkness affect the human body? Research and present your findings to your class.

Review

Summarize It!

1. **Construct** Diagram how the Sun-Earth system causes the seasonal patterns. Include illustrations and science terms in your diagram.

Three-Dimensional Thinking

2. Which best explains why Earth is colder at the poles than at the equator?

    A Earth is farther from the Sun at the poles than at the equator.

    B Earth's orbit is not a perfect circle.

    C Earth's rotation axis is tilted.

    D Earth's surface is more tilted at the poles than at the equator.

Use the figure to answer question 3.

3. Predict the pattern of the light that will appear on each sheet of paper when the flashlight is turned on.

    A The pattern on the vertical paper will be a circle, the pattern that will appear on the tilted paper will be elongated and spread out.

    B The pattern on the papers will change because the light will fluctuate.

    C The pattern that will appear on both pieces of paper is the same—a circle of light.

    D The pattern that will appear on both pieces of paper is the same—elongated and spread out.

Real-World Connection

4. Analyze How do you think seasonal change affects your local community's usage of energy?

_____

_____

_____

_____

5. Infer How do you think animals are affected by seasonal change?

_____

_____

_____

_____

 Still have questions?
Go online to check your understanding of Earth's motion and seasons.

 REVISIT PAGE KEELEY SCIENCE PROBES

Do you still agree with the friend you chose at the beginning of the lesson? Return to the Science Probe at the beginning of the lesson. Explain why you agree or disagree with that friend now.

 EXPLAIN THE PHENOMENON

Revisit your claim about what causes the seasonal changes. Explain how your evidence supports your claim.

START PLANNING

STEM Module Project
Science Challenge

Now that you learned about Earth's motion, the tilt of Earth's axis, and the seasons, go to your Module Project to begin incorporating information from this lesson into your model. You will need to be able to show how the seasonal patterns are visible in your model.

LESSON 2 LAUNCH

Phases of the Moon

Many people have different ideas about what causes us to see different parts of the Moon (moon phases). Which idea below best matches your thinking?

A.   The Earth casts a shadow on the Moon that allows us to see only the lit part.

B.   The Moon moves into the Sun's shadow, blocking out part of the Moon's light.

C.   The part we see depends on where the Moon is in relation to Earth and the Sun.

D.   Sun's movement around Earth causes different parts of the Moon to be reflected.

E.   The Moon's rotation causes different parts of the Moon to be reflected back to Earth.

F.   None of these. I think there is something else that causes us to see different moon phases.

Explain your thinking. Describe your ideas about why we see different phases of the Moon.

_____

_____

_____

You will revisit your response to the Science Probe at the end of the lesson.

Lunar Phases

ENCOUNTER
THE PHENOMENON | Why isn't the Moon always full?

Place a ball on a level surface. Position a flashlight so that the light beam shines fully on one side of the ball. Stand behind the flashlight and make a drawing of the ball's appearance in the space below.

Now stand behind the ball and sketch the ball's appearance.

Finally, stand to the left of the ball and illustrate the ball's appearance.

Moon Metamorphosis

GO ONLINE
Watch the video *Moon Metamorphosis* to see this phenomenon in action.

EXPLAIN
THE PHENOMENON

You just modeled how the Moon can appear full, only partially lit, or seem to disappear entirely. Are you starting to get some ideas about what might cause this phenomenon? Use your observations from the previous activity to make a claim about the Moon's changing appearance.

CLAIM
The Moon's changing appearance is caused by...

COLLECT EVIDENCE as you work through the lesson. Then return to these pages to record your evidence.

EVIDENCE
A. What evidence have you collected to explain from where the Moon receives its light?

MORE EVIDENCE

B. What evidence have you collected to explain the relationship between the Moon's revolution and lunar phases?

When you are finished with the lesson, review your evidence. If necessary, based on the evidence, revise your claim.

REVISED CLAIM

The Moon's changing appearance is caused by...

Finally, explain your reasoning for how and why your evidence supports your claim.

REASONING

The evidence I collected supports my claim because...

How are we able to see the Moon?

Imagine what people thousands of years ago thought when they looked up at the Moon. They might have wondered why the Moon shines and why it seems to change shape. They probably would have been surprised to learn that the Moon does not emit light at all. Unlike the Sun, the Moon is a solid object that does not emit its own light. So where does the Moon's light come from? Let's investigate!

INVESTIGATION

Foil Moon

1. Observe your teacher's demonstration and write your observations below.

2. Compare the foil-wrapped ball to the unwrapped ball and how the light appeared on each.

3. Make a claim about how these observations might explain where the Moon receives its light.

Light from the Sun The Moon does not produce or emit its own light. The Moon receives its light from the Sun. The Moon's surface is light in color and reflects the light from the Sun.

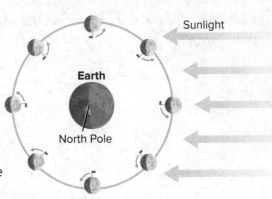

The amount of light that is reflected off the Moon's surface is actually quite low. Only about 12 percent of the light from the Sun is reflected off the Moon. That 12 percent still makes the Moon the brightest object in our night sky!

COLLECT EVIDENCE

Where does the Moon receive its light? Record your evidence (A) in the chart at the beginning of the lesson.

How does the Moon move?

While Earth is revolving around the Sun, the Moon is revolving around Earth. The gravitational pull of Earth on the Moon causes the Moon to move in an orbit around Earth. The Moon makes one revolution around Earth every 27.3 days. Does the Moon rotate like Earth? Let's find out!

> **Want more information?**
> Go online to read more about the motion of the moon.

INVESTIGATION

The Motion of the Moon

1. Choose a partner. One person represents the Moon. The other represents Earth.

2. For the first model, have the *Moon* move counterclockwise around *Earth*, always facing the same wall. *Earth* should turn counterclockwise to continuously face the *Moon*.

3. For the second model, have the *Moon* move around *Earth* always facing *Earth*. *Earth* should turn counterclockwise to continuously face the *Moon*.

4. For which model was the *Moon* both revolving and rotating?

_____

5. In each model, how many times did the *Moon* rotate during one revolution around *Earth*?

_____

_____

_____

_____

_____

THREE-DIMENSIONAL THINKING

Explain how the Moon can be rotating if the same side of the Moon is always facing Earth.

_____

_____

_____

_____

One Rotation, One Revolution The Moon also rotates as it revolves around Earth. One complete rotation of the Moon also takes 27.3 days. This means the Moon makes one rotation in the same amount of time that it makes one revolution around Earth. Because the Moon makes one rotation for each revolution of Earth, the same side of the Moon always faces Earth. This side of the Moon is called the near side. The side of the Moon that cannot be seen from Earth is called the far side of the Moon.

Why does the Moon appear to change shape?

The Sun is always shining on half of the Moon, just as the Sun is always shining on half of Earth. However, as the Moon moves around Earth, usually only part of the Moon's near side is lit. The portion of the Moon or a planet reflecting light as seen from Earth is called a **phase.** How many phases are there? Let's find out!

FOLDABLES

Go to the Foldables® library to make a Foldable® that will help you take notes while reading this lesson.

Copyright © McGraw-Hill Education

LAB Moon Phases

Safety 🧤 ✂️ ⚡ 🧪

Materials

foam ball

lamp

pencil

stool

Procedure

1. Read and complete a lab safety form.

2. Have one group member hold a foam ball that represents the Moon. Make a handle for the ball by inserting a pencil about two inches into the foam ball. Another group member will represent an observer on Earth. Have the observer sit on a stool and record observations during the activity.

3. Place a lamp on a desk or other flat surface. Remove the shade from the lamp. The lamp represents the Sun.

4. Turn on the lamp and darken the lights in the room.

CAUTION: Do not touch the bulb or look directly at it after the lamp is turned on.

5. Position the Earth observer's stool about 1 m from the Sun. Position the Moon 0.5–1 m from the observer so that the Sun, Earth, and the Moon are in a line. The student holding the Moon should hold the Moon so it is completely illuminated on one half. The observer records the phase and what the phase looks like in the data table on the next page.

6. Move the Moon counterclockwise about one-eighth of the way around its "orbit" of Earth. The observer swivels on the stool to face the Moon and records the phase.

7. Continue the Moon's orbit until the Earth observer has recorded all the Moon's phases.

8. Return to your positions as the Moon and Earth observer. Choose a part in the Moon's orbit that you did not model. Predict what the Moon would look like in that position, and check if your prediction is correct.

9. Follow your teacher's instructions for proper clean up.

Data and Observations

Position	Phase	Description
1		
2		
3		
4		
5		
6		
7		
8		

Analyze and Conclude

10. Use your observations to explain how the positions of the Sun, the Moon, and Earth produce the different phases of the Moon.

_____

_____

_____

_____

_____

_____

11. Why is half of the Moon always lit? Why do you usually see only part of the Moon's lit half?

_____

_____

_____

_____

_____

12. Based on your observations, why is the Moon not visible from Earth during the new moon phase?

_____

_____

Lunar Phases While the Moon rotates and revolves around Earth in 27.3 days, the length of the lunar phases cycle is 29.5 days. The Moon's appearance changes as Earth and the Moon move. Depending on where the Moon is in relation to Earth and the Sun, observers on Earth see only part of the light the Moon reflects from the Sun.

Waxing Phases During the waxing phases, more of the Moon's near side is lit each night.

Week 1—First Quarter As the lunar cycle begins, a sliver of light can be seen on the Moon's western edge. Gradually the lit part becomes larger. By the end of the first week, the Moon is at its first quarter phase. In this phase, the Moon's entire western half is lit.

First Quarter Moon

Week 2—Full Moon During the second week, more and more of the near side becomes lit. When the Moon's near side is completely lit, it is at the full moon phase.

Full Moon

Waning Phases During the waning phases, less of the Moon's near side is lit each night. As seen from Earth, the lit part is now on the Moon's eastern side.

Week 3—Third Quarter During this week, the lit part of the Moon becomes smaller until only the eastern half of the Moon is lit. This is the third quarter phase.

Third Quarter Moon

Week 4—New Moon During this week, less and less of the near side is lit. When the Moon's near side is completely dark, it is at the new moon phase.

You've just read that as the Moon revolves around Earth, the part of the Moon's near side that is lit changes. Let's examine what the Moon looks like at different places in its orbit.

New Moon

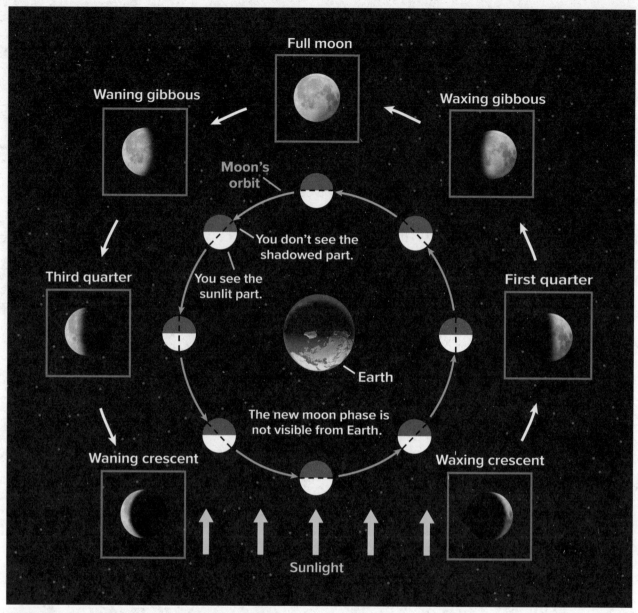

The Moon's motion around Earth causes the Moon to rise, on average, about 50 minutes later each day. The figure below shows how the Moon looks at midnight during three phases of the lunar cycle.

First quarter

At midnight, the first quarter moon is setting. It rises during the day at about noon.

Full moon

The full moon is highest in the sky at about midnight. It rises at sunset and sets at sunrise.

Third quarter

The third quarter moon rises at about midnight, about six hours later than the full moon rises.

COLLECT EVIDENCE

How does the Moon's revolution contribute to lunar phases? Record your evidence (B) in the chart at the beginning of the lesson.

Return to the Moon

Exploring Earth's Moon is a step toward exploring other planets and building outposts in space.

The United States undertook a series of human spaceflight missions from 1961–1975 called the Apollo program. The goal of the program was to land humans on the Moon and bring them safely back to Earth. Six of the missions reached this goal. The Apollo program was a huge success, but it was just the beginning.

NASA began another space program that had a goal to return astronauts to the Moon to live and work. However, before that could happen, scientists needed to know more about conditions on the Moon and what materials are available there. Collecting data was the first step. In 2009, NASA launched the Lunar Reconnaissance Orbiter (LRO) spacecraft. The LRO is in an elliptical orbit that takes it over both of the Moon's poles. As it orbits, it collects detailed data that scientists can use to make maps of the Moon's features and resources, such as deep craters that formed on the Moon when comets and asteroids slammed into it billions of years ago. Some scientists predicted that these deep craters contain frozen water.

One of the instruments launched with the LRO was the Lunar Crater Observation and Sensing Satellite (LCROSS). LCROSS observations confirmed the scientists' predictions that water exists on the Moon. A rocket launched from LCROSS impacted the Cabeus crater near the Moon's south pole. The material that was ejected after the rocket's impact included water. NASA's goal of returning astronauts to the Moon was delayed, and their missions now focus on exploring Mars instead. But the discoveries made on the Moon will help scientists develop future missions that could take humans farther into the solar system.

Apollo SPACE PROGRAM

The Apollo Space Program included 17 missions. Here are some milestones:

May 25 1961
President John F. Kennedy announces the goal of landing a man on the Moon by the end of the decade.

December 21–27 1968
Apollo 8 First manned spacecraft orbits the Moon.

July 16–24 1969
Apollo 11 First humans, Neil Armstrong and Buzz Aldrin, walk on the Moon.

July 1971
Apollo 15 Astronauts drive the first rover on the Moon.

December 7–19 1972
Apollo 17 The first phase of the Moon ended with the last lunar landing mission.

It's Your Turn

Write a paragraph about whether or not you think humans should establish an outpost on the Moon and the importance of water to this effort.

Review

Summarize it!

1. **Identify** the moon phase represented by each illustration. Then draw what each phase looks like from Earth.

 = Sun   = Moon   = Earth

Three-Dimensional Thinking

2. A new moon occurs once every 29.5 days. Why must the Sun, Earth, and the Moon be aligned in order for the new moon to occur?

    A   No sunlight is reflected off Earth at this point.

    B   Sunlight directed toward Earth is blocked by the Moon.

    C   The Moon does not orbit in the same plane as Earth.

    D   The Moon is not directing any light toward Earth at this point.

Use the image to answer questions 3 and 4.

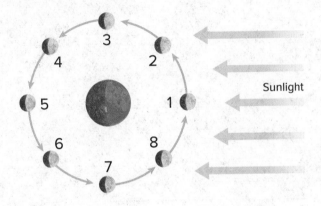

3. Predict the locations when the Moon is in a waxing phase.

    A   1, 5

    B   2, 3, 4

    C   7, 8

    D   3, 7

4. Predict which location is the phase seen from Earth at the end of the second week of the lunar cycle.

    A   1

    B   3

    C   5

    D   7

Real-World Connection

5. **Evaluate** Would the phases of the Moon be affected if the Moon did not make one rotation for each revolution of Earth? Explain.

_____

_____

_____

_____

6. **Infer** Imagine the Moon rotates twice in the same amount of time the Moon orbits Earth once. Would you be able to see the Moon's far side from Earth? Explain your reasoning.

_____

_____

_____

_____

_____

 Still have questions?
Go online to check your understanding of lunar phases.

 REVISIT
PAGE KEELEY
SCIENCE PROBES

Do you still agree with the statement you chose at the beginning of the lesson? Return to the Science Probe at the beginning of the lesson. Explain why you agree or disagree with that statement now.

KEEP PLANNING
STEM Module Project
Science Challenge

Now that you learned about the lunar phases, go to your Module Project to continue planning your model. Keep in mind that you want to use your model to help explain the patterns of the lunar phases.

 EXPLAIN
THE PHENOMENON

Revisit your claim about what causes the lunar phases. Explain how your evidence supports your claim.

Eclipses

Six students were discussing what causes eclipses. This is what they said:

Isabella: I think an eclipse of the Moon happens about every 29 days.

Sam: I think the Sun stops giving off light during an eclipse.

June: I think we can't see the Moon when Earth's shadow falls on it.

Olivia: I think we can't see the Sun when Earth's shadow falls on it.

Santiago: I think each new moon is the same as an eclipse.

Mateo: I think something else causes eclipses to occur.

Which student do you agree with the most? Explain why you agree with that student.

_____

_____

_____

_____

_____

_____

You will revisit your response to the Science Probe at the end of the lesson.

Eclipses

ENCOUNTER
THE PHENOMENON

What is causing these "bites" out of the Sun?

GO ONLINE Watch the video *Eclipses* to see this phenomenon in action.

Record your observations about the phenomena in the space provided. Why do you think these phenomena occur?

EXPLAIN
THE PHENOMENON

You just watched eclipses of the Sun and Moon. Are you starting to get some ideas about what is happening in the Sun-Earth-Moon system during each type of eclipse? Use your observations about the phenomena to make a claim about how eclipses occur.

CLAIM

Eclipses occur because...

 COLLECT EVIDENCE as you work through the lesson. Then return to these pages to record your evidence.

EVIDENCE

A. What evidence have you discovered to explain how solar eclipses occur?

B. What evidence have you discovered to explain how lunar eclipses occur?

MORE EVIDENCE

C. What evidence have you discovered to explain the impact of the Moon's tilted orbit on eclipse formation?

When you are finished with the lesson, review your evidence. If necessary, based on the evidence, revise your claim.

REVISED CLAIM

Eclipses occur because...

Finally, explain your reasoning for how and why your evidence supports your claim.

REASONING

The evidence I collected supports my claim because...

What makes a shadow?

A shadow results when one object blocks the light that another object emits or reflects. For example, when a tree blocks light from the Sun, it casts a shadow. If you want to stand in the shadow of a tree, the tree must be in a line between you and the Sun.

Each object in the solar system creates a shadow as it blocks the path of the Sun's light. What happens to an object's shadow when the object moves? Let's find out!

 Want more information?
Go online to read more about solar and lunar eclipses.

FOLDABLES

Go to the Foldables® library to make a Foldable® that will help you take notes while reading this lesson.

LAB Beyond a Shadow of a Doubt

Safety

Materials

flashlight

object

Procedure

1. Read and complete a lab safety form.

2. Select an object provided by your teacher.

3. Shine a flashlight on the object, projecting its shadow on the wall.

4. While holding the flashlight in the same position, move the object closer to the wall—away from the light. Then, move the object toward the light. Record your observations in the Data and Observations section on the next page.

5. Follow your teacher's instructions for proper cleanup.

Data and Observations

Analyze and Conclude

6. Compare and contrast the shadows created in each situation. Describe the shadows and how they changed.

_____

_____

_____

_____

_____

7. Imagine you could look at the flashlight from behind your object, looking from the darkest and lightest parts of the object's shadow. How much of the light source do you think you could see from each location?

_____

_____

_____

_____

The Umbra and the Penumbra If you go outside on a sunny day and look carefully at a shadow on the ground, you might notice that the edges of the shadow are not as dark as the rest of the shadow. Light from the Sun and other sources casts shadows with two distinct parts, as shown in the figure below.

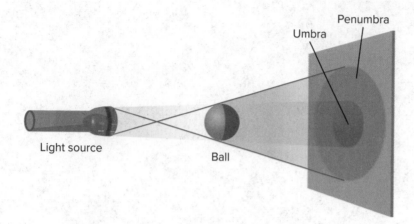

- The **umbra** is the central, darker part of a shadow where light is totally blocked. If you stood within an object's umbra, you would not see the light source at all.

- The **penumbra** is the lighter part of a shadow where light is partially blocked. If you stood within an object's penumbra, you would be able to see only part of the light source.

What is a solar eclipse?

As the Sun shines on the Moon, the Moon casts a shadow that extends out into space. Sometimes the Moon passes between Earth and the Sun. This can only happen during the new moon phase. Let's investigate!

LAB Casting Shadows

Safety

Materials

light source
large foam ball
small foam ball
pencil

Procedure

1. Read and complete a lab safety form.

2. Working with your partner, use a pencil to connect two foam balls.

3. While one person holds the foam balls, the other should stand 1 m away and shine a light on the foam balls. The foam balls and light should be in a direct line, with the smallest foam ball closest to the light.

4. Sketch and describe your observations in the Data and Observations section.

5. Follow your teacher's instructions for proper cleanup.

Data and Observations

Analyze and Conclude

6. Explain the relationship between the two types of shadows and solar eclipses.

_____

_____

_____

_____

_____

7. In the lab you used two different-sized foam balls. Which did each of them represent? Now that you know what a solar eclipse is, explain what a lunar eclipse is using words and illustrations.

8. Using what you have learned in this lab, predict how a lunar eclipse occurs.

_____

_____

_____

_____

_____

_____

Solar Eclipse As the Sun shines on the Moon, the Moon casts a shadow that extends out into space. Sometimes the Moon passes between Earth and the Sun. Solar eclipses can only occur during the new moon phase. When Earth, the Moon, and the Sun are lined up, the Moon casts a shadow on Earth's surface. When the Moon's shadow appears on Earth's surface, a **solar eclipse** is occurring. As Earth rotates, the Moon's shadow moves along Earth's surface. The type of eclipse you see—total or partial—depends on where you are in the path of the eclipse.

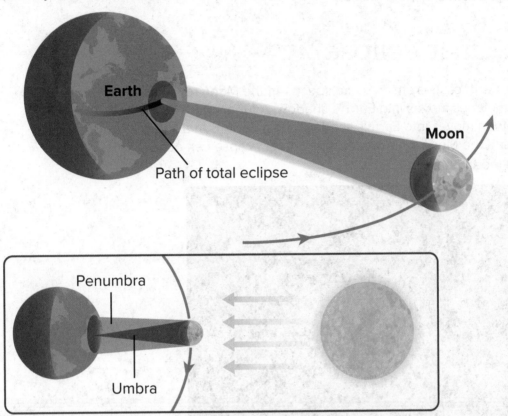

Earth

Moon

Path of total eclipse

Penumbra

Umbra

Total Solar Eclipses You can only see a total solar eclipse from within the Moon's umbra. During a total solar eclipse, the Moon appears to cover the Sun completely, as shown in the figure below. Then, the sky becomes dark enough that you can see stars. A total solar eclipse lasts no longer than about seven minutes.

The Sun's Changing Appearance During a Total Solar Eclipse

The Motion of the Moon in the Sky During a Total Solar Eclipse

Sun   Moon

Moon's motion

Partial Solar Eclipses You can see a partial solar eclipse from within the Moon's much larger penumbra. The stages of a partial solar eclipse are similar to the stages of a total solar eclipse, except that the Moon never completely covers the Sun.

COLLECT EVIDENCE

What is necessary for a solar eclipse to take place? Record your evidence (A) in the chart at the beginning of the lesson.

What is a lunar eclipse?

Just like the Moon, Earth casts a shadow into space. As the Moon revolves around Earth, it sometimes moves into Earth's shadow. A **lunar eclipse** occurs when the Moon moves into Earth's shadow. Then Earth is in a line between the Sun and the Moon. This means that a lunar eclipse can occur only during the full moon phase.

Like the Moon's shadow, Earth's shadow has an umbra and a penumbra. Different types of lunar eclipses occur depending on which part of Earth's shadow the Moon moves through. Unlike solar eclipses, you can see any lunar eclipse from any location on the side of Earth facing the Moon.

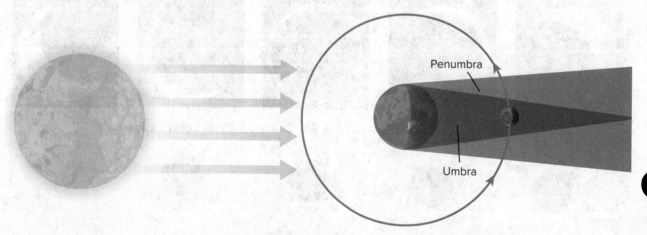

Penumbra

Umbra

Total Lunar Eclipses When the entire Moon moves through Earth's umbra, a total lunar eclipse occurs. The Moon's appearance changes as it gradually moves into Earth's penumbra, then into Earth's umbra, back into Earth's penumbra, and then out of Earth's shadow entirely.

THREE-DIMENSIONAL THINKING

You've just learned that during a total lunar eclipse, the entire Moon passes through Earth's umbra. The Moon gradually darkens until a dark shadow covers it completely. **Construct an explanation** to describe how a total lunar eclipse would look compared with a total solar eclipse.

_____

_____

_____

You can still see the Moon even within Earth's umbra. Although Earth blocks most of the Sun's rays, Earth's atmosphere deflects some sunlight into Earth's umbra. This is also why you can often see the unlit portion of the Moon on a clear night. This reflected light has a reddish color and gives the Moon a reddish tint during a total lunar eclipse.

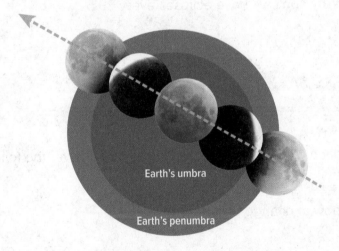

Earth's umbra

Earth's penumbra

Partial Lunar Eclipses When only part of the Moon passes through Earth's umbra, a partial lunar eclipse occurs. The stages of a partial lunar eclipse are similar to those of a total lunar eclipse, except the Moon is never completely covered by Earth's umbra. The part of the Moon in Earth's penumbra appears only slightly darker, while the part of the Moon in Earth's umbra appears much darker.

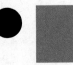

COLLECT EVIDENCE

What is necessary for a lunar eclipse to take place? Record your evidence (B) in the chart at the beginning of the lesson.

THREE-DIMENSIONAL THINKING

What patterns did you notice between solar and lunar eclipses? How do those patterns impact who can see each type of eclipse? Explain your reasoning.

_____

_____

_____

_____

_____

_____

Why aren't there eclipses every month?

A solar eclipse can only happen during the new moon phase. Lunar eclipses can only occur during a full moon. You have learned that each of those phases occurs every 29.5 days. Why don't we have eclipses every 29.5 days? Let's find out!

INVESTIGATION

Eclipse Essentials

GO ONLINE Watch the animation _To Eclipse or Not to Eclipse_ and answer the following questions.

1. What causes the Moon's shadow to change placement?

_____

_____

_____

_____

2. From the video, what must happen for a solar eclipse to occur?

_____

_____

3. Why don't we see lunar eclipses every month?

_____

_____

_____

_____

Rare Eclipses Lunar eclipses can only occur during a full moon phase, when the Moon and the Sun are on opposite sides of Earth. However, lunar eclipses do not occur during every full moon because of the tilt of the Moon's orbit with respect to Earth's orbit. During most full moons, the Moon is slightly above or slightly below Earth's penumbra.

Solar eclipses can occur only during a new moon, when Earth and the Sun are on opposite sides of the Moon. However, solar eclipses do not occur during every new moon phase. The figure below shows why. The Moon's orbit is tilted slightly compared to Earth's orbit. As a result, during most new moons, Earth is either above or below the Moon's shadow. However, every so often the Moon is in a line between the Sun and Earth. Then the Moon's shadow passes over Earth and a solar eclipse occurs.

A solar eclipse occurs when the Moon crosses directly between the Sun and Earth.

No solar eclipse occurs when the Moon is not directly between the Sun and Earth.

COLLECT EVIDENCE
How does the Moon's tilted orbit affect eclipses? Record your evidence (C) in the chart at the beginning of the lesson.

THREE-DIMENSIONAL THINKING

You've learned that lunar and solar eclipses do not occur every month. But scientists have been able to detect **patterns** in the Sun–Earth–Moon system and can predict when the next eclipses will occur. Locate an eclipse calendar on the Internet. After examining the calendar and locations, **construct an explanation** of why most people will not witness a total solar eclipse in their lifetime.

_____

_____

_____

_____

_____

_____

What is the next eclipse you may be able to witness? When will the eclipse occur?

_____

_____

_____

_____

_____

GO ONLINE for an additional opportunity to explore!

PHYSICAL SCIENCE Connection Want to know more about eclipses? Investigate eclipses and their applications by performing the following activity.

☐ **Investigate** the relationship between the theory of relativity and solar eclipses in the **Scientific Text** _Testing Einstein's Theory of Gravity._

A Closer Look: Solar Eclipse Eye Safety

The closest star to Earth is the Sun. The Sun is a star you can observe during the day. Looking directly at the Sun, even for a moment, can be very harmful to your eyes. But is it safe to look at the Sun during a solar eclipse while the Moon blocks the sunlight?

Although the Sun's visible light is blocked by the Moon during a solar eclipse, ultraviolet (UV) rays are still being emitted and could cause damage to your eyes. If you want to observe a solar eclipse, special precautions should be used to stay safe.

To safely observe a solar eclipse, you should use protective eyewear, such as eclipse glasses, telescope filters, or welder's glasses. These items have a thin layer of aluminum, chromium, or silver on the surface that reduces the effects of UV, visible, and infrared light.

It's Your Turn

Research and Present Go online to research when the next solar eclipse will occur. Create a blog to promote safety awareness for viewing a solar eclipse properly.

Review

Summarize It!

1. Illustrate the positions of the Sun, Earth, and the Moon during a solar eclipse and during a lunar eclipse. Also identify the correct phase that the Moon is in during each type of eclipse.

Solar eclipse

Lunar eclipse

Three-Dimensional Thinking

Use the figure to answer questions 2 and 3.

2. What is occurring in this figure?

   A lunar eclipse

   B lunar and solar eclipse

   C partial lunar eclipse

   D solar eclipse

3. Why would someone in North America not be able to view this eclipse?

   A Because this is a lunar eclipse and Earth casts a very small shadow.

   B Because this is a lunar eclipse and North America is experiencing day.

   C Because this is a solar eclipse and North America is experiencing night.

   D Because this is a solar eclipse and the Moon casts a very small shadow.

4. Where do you have to be located to be able to see a total eclipse?

   A anywhere on the continent where the eclipse is occuring

   B within the penumbra

   C within the penumbra and umbra

   D within the umbra

Real-World Connection

5. Infer During a lunar eclipse, Earth casts a shadow on the Moon. What type of shadow would Earth cast if it were flat? What type of shadow does Earth cast on the Moon during a lunar eclipse?

_____

_____

_____

_____

6. Predict Many communities rely on solar energy for daily life uses. Suppose the Moon was twice its size and cast a bigger shadow on Earth during a solar eclipse, causing the light from the Sun to be blocked for days. How would this affect the communities who rely on solar energy?

_____

_____

_____

_____

 Still have questions?
Go online to check your understanding of solar and lunar eclipses.

REVISIT **SCIENCE PROBES** Do you still agree with the student you chose at the beginning of the lesson? Return to the Science Probe at the beginning of the lesson. Explain why you agree or disagree with that student now.

PLAN AND PRESENT
STEM Module Project
Science Challenge
Now that you learned about the eclipses, go to your Module Project to finish your model. Keep in mind that your model needs to correctly represent the positions of the Sun, Earth, and the Moon when describing eclipses.

EXPLAIN
THE PHENOMENON

Revisit your claim about what causes eclipses. Explain how your evidence supports your claim.

Patterns in the Sky

Imagine you are an astronomer who works for a science museum. A group of third-grade students is visiting the museum to learn more about the Sun–Earth–Moon system.

You have been asked to develop a model to help students visualize the patterns of lunar phases, eclipses of the Sun and the Moon, and seasons. To test your model, you will use it to predict different patterns in the sky.

Planning After Lesson 1

In your model, how will you show the distance between Earth and the Sun? How will you show the position of Earth's axis as it orbits the Sun?

Research the different ways you can build and present your models.

Planning After Lesson 1, continued

In the space below, sketch a model showing why seasons occur. Label all components of your model, and include captions explaining Earth's movements and how they are related to seasons. Also describe how sunlight travels and hits different parts of Earth during each season.

Planning After Lesson 2

In your model, how will you show the movements of the Moon and the Earth–Moon system?

Sketch your model of lunar phases in the space below. Label all components of your model, and include captions explaining why the pattern of lunar phases occurs.

Planning After Lesson 3

In your model, how will you show why solar and lunar eclipses do not occur every month?

Sketch your model of a solar eclipse and a lunar eclipse in the space below. Label all components of your model, and include captions explaining why eclipses occur.

Develop Your Model

Look over the materials provided by your teacher. Then, as a group, design your model. List the materials you will use and each step you will take to build your model. Describe how you will address proportional relationships (size and distance) in your model.

Evaluate Your Model

Once your model is complete, make and record observations of seasons, lunar phases, and solar and lunar eclipses. Then, identify the model elements in the table below.

Model Elements	Descriptions
Components (What are the different parts of my model?)	
Relationships (How do the components of my model interact?)	
Connections (How does my model help me understand the phenomenon?)	

Make Predictions

Use your model to make predictions based on the descriptions below.

Description	Prediction
The Sun, Earth, and the Moon are lined up in a straight row with the Moon in the middle. Predict the phase of the Moon.	
The moon is in the last quarter phase. Predict the relative positions of Earth, the Sun, and the Moon.	
The Sun, Earth, and the Moon form a right angle. Predict whether an eclipse will occur. Explain your prediction.	
A solar eclipse is occurring. Predict the relative positions of the Sun, Earth, and the Moon.	
Earth's southern axis is tilted toward the Sun. Predict the seasons in the southern and northern hemispheres.	
It is summer in the northern hemisphere. Predict the position of Earth in relation to the Sun.	

Present Your Model

Present your model before the class.

Is your model appropriate for a third-grade audience? If yes, explain why. If no, explain how you could make it more appropriate.

Why is it difficult to show scale size or scale distance in a model of the Sun-Earth-Moon system?

In addition to problems with scale, what are some limitations of models? What are some advantages?

Congratulations! You've completed the Science Challenge requirements.

Module Wrap-Up

● **REVISIT**
THE PHENOMENON

Using the model you developed for the Science Challenge, explain how the Sun, Earth, and the Moon cause seasons, eclipses, and Moon phases.

OPEN INQUIRY

If you had to ask one question about what you studied, what would it be?

_____

_____

_____

_____

Plan and conduct an investigation to answer this question.

Exploring the Universe

ENCOUNTER
THE PHENOMENON

What makes up the universe, and how does gravity affect the universe?

Viewing the Universe

GO ONLINE
Watch the video *Viewing the Universe* to see this phenomenon in action.

Collaborate A system is an organized group of components, or objects, that together form a whole. The universe is a system. With your class, develop a list of components that make up this system. Are there any relationships among the components? How does gravity impact the components? Record your thoughts in the space below.

Wanted: Space Investigator

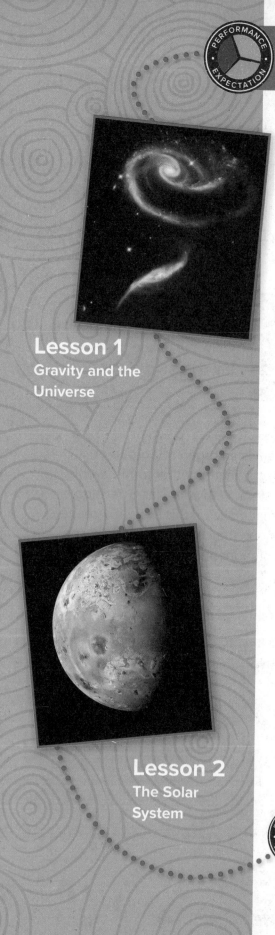

Lesson 1
Gravity and the Universe

You receive an email with the subject line "Wanted: Space Investigators." The email reads:

"The National Aeronautics and Space Administration (NASA) has selected you to manage a team of space investigators to analyze and model objects in space. Your team will deliver a presentation to government officials who want to know more about our galaxy and solar system, and the technology used to explore space."

In 3 ... 2 ... 1 ... begin your mission, Space Investigator!

Start Thinking About It

What properties would you include in a scale model of the solar system?

Lesson 2
The Solar System

STEM Module Project

Planning and Completing the Science Challenge How will you meet this goal? The concepts you will learn throughout this module will help you plan and complete the Science Challenge. Just follow the prompts at the end of each lesson!

Gravity in Space?

Olivia asked her friends what they know about gravity in space. This is what they said:

Justin: Gravity is a force that does not exist in space because objects float in space.

Taylor: I think gravity is a force that only exists on the surfaces of planets.

Juanita: Gravity is a force that exists between all objects that have mass— even in space!

Which friend do you agree with most? Explain why you agree.

_____

_____

_____

_____

_____

_____

You will revisit your response to the Science Probe at the end of the lesson.

LESSON 1
Gravity and the Universe

ENCOUNTER THE PHENOMENON | What would space be like without gravity?

GO ONLINE

Watch the video *Forming Galaxies* to see this phenomenon in action.

As you watch the video, refer back to the list of components in the universe that you compiled earlier. Add any new objects to the list as needed. After the video is over, discuss with a partner what you think would happen to these components if gravity did not exist.

EXPLAIN
THE PHENOMENON

Are you starting to get some ideas about how gravity impacts components in solar and galaxy systems? Use your observations about the phenomenon to make a claim about the role of gravity in the universe.

CLAIM

Gravity plays a vital role in the universe because...

 COLLECT EVIDENCE as you work through the lesson. Then return to these pages to record your evidence.

EVIDENCE

A. What evidence have you discovered to explain the factors that affect gravitational force?

B. What evidence have you discovered to explain how the solar system formed?

MORE EVIDENCE

C. What evidence have you discovered to explain how gravity impacts the formation and structure of galaxies?

When you are finished with the lesson, review your evidence. If necessary, based on the evidence, revise your claim.

REVISED CLAIM

Gravity plays a vital role in the universe because...

Finally, explain your reasoning for how and why your evidence supports your claim.

REASONING

The evidence I collected supports my claim because...

Copyright © McGraw-Hill Education

What is gravity?

Have you ever dropped your books while walking to your classroom? Or bounced a basketball in gym class? What keeps you, your books, a basketball, and everything else on Earth... on Earth?

What Goes Up Must Come Down

Study the figure below. In models of gravitational force, strength is represented by the thickness of the arrows. The thicker the arrows, the greater the force.

The two objects in row A are the same distance apart as the two objects in row B. One of the objects in row B has more mass, creating a stronger gravitational force between the two objects in row B.

All four objects have the same mass. The two objects in row C are closer to each other than the two objects in row D and, therefore, have a stronger gravitational force between them.

1. What factors affect the strength of the gravitational force between objects?

_____

_____

2. Use the model above to construct an argument on why Earth exerts a greater gravitational force on you than other objects do.

_____

_____

_____

Want more information?
Go online to read more about the role of gravity in the universe.

FOLDABLES
Go to the Foldables® library to make a Foldable® that will help you take notes while reading this lesson.

The Force of Gravity Objects fall to the ground because Earth exerts an attractive force on them. **Gravity** is an attractive force that exists between all objects that have mass. The force of gravity between two objects depends on the objects' masses and the distance between them. The more mass either object has, or the closer together they are, the stronger the gravitational force.

Earth's gravity holds us on Earth's surface. Since Earth has more mass than any object near you, it exerts a greater gravitational force on you than other objects do. You don't notice the gravitational force between less massive objects.

 THREE-DIMENSIONAL THINKING

1. Using what you have learned, create a diagram **modeling** why the Moon orbits Earth and not the Sun. Use force arrows to show gravitational relationships within the **system.**

2. Using your model as evidence, **argue** how gravitational force is dependent upon the masses of interacting objects and the distance between them. Construct your argument in your Science Notebook.

COLLECT EVIDENCE
What factors affect gravitational force? Record your evidence (A) in the chart at the beginning of the lesson.

What is gravity's role in the formation of stars?

Just as the force of gravity pulls a basketball to the floor, the force also pulls dust and gas inward in space. What impact does this have on objects in the universe, such as galaxies and stars? Let's find out.

PHYSICAL ▶ **Connection** When gas and dust clump together and form nebulae, the particles bump into each other and move faster. Gravity makes these nebulae hotter—as the particles are pulled closer together as a result of gravity, they start to bump into each and cause them to move faster, which causes the temperature to increase. The hotter something is, the more quickly its atoms move. As atoms move, they collide. If a gas is hot enough and its atoms move quickly enough, the nuclei of some of the atoms combine. Nuclear fusion is a process that occurs when the nuclei of several atoms combine into one larger nucleus. Once nuclear fusion has occurred, a star is formed.

What is the role of gravity in the formation of the solar system?

When the solar system formed, it was an enormous ball of gas and dust spinning slowly in space. As gravity pulled it closer together, the solar system spun faster. What happened to the shape of the solar system as it spun faster?

 LAB Changing Shape

Safety

Materials

salt dough

small bucket

stopwatch

string

Procedure

1. Read and complete a lab safety form.

2. Make a round ball about the size of your fist from a piece of salt dough.

3. Place the dough in a small bucket. Attach 1 m of sturdy string to the bucket's handle. **CAUTION:** *Be sure the string is securely attached.*

4. **CAUTION:** *Stand away from all furniture and people.* Whirl the bucket around your head for 1 min.

5. Lower the bucket. Observe the salt dough and record your observations in the Data and Observations section.

Data and Observations

Analyze and Conclude

6. What happened to the salt dough? What other objects change shape as they spin?

_____

_____

_____

_____

_____

7. How do you think gravity influenced the shape of the early solar system?

_____

_____

_____

_____

_____

_____

Solar System Formation and Structure The solar system formed from a rotating cloud of gas, ice, and dust called a nebula. Gravity caused the cloud to collapse inward. Because the cloud was still rotating after the collapse, it flattened into a disk. Gravity pulled gas and dust toward the center of the disk, forming the Sun. Next, the planets began to form as gravity pulled these small particles together. As they collided, they stuck to each other and formed larger, unevenly shaped objects.

These larger objects had more mass and attracted more particles. Eventually enough matter collected and formed Earth and the seven other planets.

The leftover larger particles formed into moons that orbit the planets. Asteroids orbit the Sun in the asteroid belt, while meteoroids, dwarf planets, and comets orbit the Sun in their own paths through the solar system.

Solar nebula

Disk forms

Planets form

Solar system today

COLLECT EVIDENCE
How did the solar system form? Record your evidence (B) in the chart at the beginning of the lesson.

THREE-DIMENSIONAL THINKING
Using what you know about how the solar system formed, **construct an explanation** of how other solar systems in other galaxies form.

_____

_____

_____

_____

_____

_____

History from Space

Meteorites give a peek back in time.

More than 4.5 billion years ago, the planets of our solar system did not exist. Instead, a large disk of gas and dust, called the solar nebula, swirled around a young Sun. How did the planets and other objects in our solar system form from the solar nebula?

Denton Ebel, a cosmochemist at the American Museum of Natural History in New York City, is investigating this question. Ebel wants to know the processes by which the first solid objects formed and how they combined to form larger bodies in our solar system. He looks for clues in certain kinds of meteorites called chondrites. Chondrites contain chondrules, clumps of dust grains that melted and solidified forming small glassy beads containing crystals. Chondrites also contain calcium-aluminum inclusions (called CAIs), the oldest rocks. Chondrules and CAIs are some of the oldest material in our solar system. Combined with a fine-grained mineral matrix, they make chondrites whose compositions are similar to that of the Sun but without the gases.

▲ **Denton Ebel holds a meteorite that was ejected from the asteroid 4 Vesta by a giant impact.**

Ebel uses scientific instruments, such as a CAT-scan imager and an electron microprobe, to study chondrites. The instruments help him identify the abundance and composition of minerals in the chondrules. Using this data Ebel develops computer models based on the laws of physics and chemistry that simulate the conditions under which the chondrules, CAIs, and chondrites formed.

Through his research and collaboration with astrophysicists, Ebel can begin to understand how the solid precursors to planets formed, where in the solar nebula they formed, and why the rocky inner planets: Mercury, Venus, Mars and Earth, are different in composition from each other.

Nebular Theory

The solar system formed in stages.

- Gravity caused the solar nebula to collapse forming a rotating disk with the Sun at its center.

- As the disk cooled, rocky material condensed in the hot inner regions. Ices condensed from gas in the cooler outer regions.

- Tiny particles collided and accreted rapidly. Through repeated collisions they eventually formed the planets. Chondrites are leftovers of this process.

▲ These overlaid element maps of a single chondrule, taken with an electron microprobe, reveal magnesium-rich minerals in glass: red for magnesium, green for calcium, and blue for aluminum. Metal is black.

It's Your Turn

TIME LINE Work in groups. Learn more about the history of Earth from its formation until life began to appear. Create a time line showing major events. Present your time line to the class.

How does gravity affect objects that orbit the Sun?

Have you ever swung a ball on the end of a string in a circle over your head? In some ways, the motion of a planet around the Sun is like the motion of that ball. As shown in the figure below, the Sun's gravitational force pulls each planet toward the Sun. This force is similar to the pull of the string that keeps the ball moving in a circle. The Sun's gravitational force pulls on each planet and keeps it moving along a curved path around the Sun.

The planet would move in a straight line without the Sun's gravitational pull.

Sun

Planet

The gravitational pull of the Sun always is toward the Sun.

Elliptical orbit

Unlike the ball on the string, planets do not move in circles. Instead, they move in an elliptical shape. Let's investigate those orbits.

LAB Elliptical Orbits

Safety ✂

Materials

paper	string
cardboard	metric ruler
push pins (2)	scissors

Procedure

1. Read and complete a lab safety form.

2. Place a sheet of paper on a piece of cardboard. Insert two push pins 8 cm apart in the center of the paper.

3. Use scissors to cut a 24-cm piece of string. Tie the ends of the string together.

4. Place the loop of string around the pins. Use a pencil to draw an ellipse.

5. Measure the maximum width and length of the ellipse. Record your data in the Data and Observations section.

6. Move one of the push pins so that the pins are 5 cm apart. Repeat steps 4 and 5. Record your observations.

Data and Observations

Analyze and Conclude

7. Compare and contrast the two ellipses.

_____

_____

_____

_____

Planetary Orbits Earth was once thought to be the center of our solar system. In this geocentric model, the Sun, the Moon, and the planets revolved in circular orbits around a stationary Earth. In the early 1500s, Nicholas Copernicus proposed that Earth and other planets revolve in circular orbits around a stationary Sun, a heliocentric model.

HISTORY Connection In the 1600s, Johannes Kepler discovered that planets' orbits are ellipses, not circles. An ellipse contains two fixed points, called foci (singular, focus). Foci are equal distance from the ellipse's center and determine its shape. The Sun is at one focus. As a planet revolves, the distance between the planet and the Sun changes. Kepler also discovered that a planet's speed increases as it gets nearer to the Sun.

Why don't all celestial objects in the solar system orbit the Sun? Remember the gravitational force between objects depends on mass and distance. So, the Moon orbits Earth because the gravitational attraction between the Moon and Earth is stronger than the attraction between the Moon and the Sun.

What are galaxies?

Most people live in towns or cities where houses are close together. Not many houses are found in the wilderness. Similarly, most stars exist in galaxies. **Galaxies** are huge collections of gas, dust, and stars held together by gravity. The universe contains hundreds of billions of galaxies, and each galaxy can contain hundreds of billions of stars. And many of those stars have planets and other celestial objects orbiting them.

Groups of Galaxies Galaxies are not distributed evenly in the universe. Gravity holds them together in groups called clusters. Some clusters of galaxies are enormous. The Virgo Cluster contains about 2,000 galaxies. Most clusters exist in even larger structures called superclusters.

The Milky Way One of those galaxies is our galaxy—the Milky Way. Our solar system and many other solar systems are in the Milky Way, a galaxy that contains almost 200 billion stars. These stars and solar systems orbit around the center of the Milky Way. The Milky Way is a member of the Local Group, a cluster of about 30 galaxies.

Virgo Cluster

How are galaxies classified?

But are all galaxies just like the Milky Way? No, each galaxy is different from every other galaxy—no two galaxies are alike. Let's find out how they are different.

INVESTIGATION

Classification of Galaxies

GO ONLINE Watch the *Types of Galaxies* video to learn more about galaxies in the universe.

While watching the video, take notes in your Science Notebook of what you see.

How would you classify the galaxies you observed in the video?

_____

_____

_____

Types of Galaxies In the video you observed three different types of galaxies—elliptical, spiral, and irregular. They are classified based on their shape.

Types of Galaxies	
	Spiral Galaxies The stars, gas, and dust in a spiral galaxy exist in the spiral arms that begin at the central disk. Some spiral arms are long and symmetric; others are short and stubby. Spiral galaxies are thicker near the center, a region called the central bulge. A spherical halo of globular clusters and older, redder stars surrounds the disk.
Elliptical Galaxies Unlike spiral galaxies, elliptical galaxies do not have spiral arms. Elliptical galaxies have a higher percentage of old, red stars. They contain little or no gas. Scientists suspect that many elliptical galaxies form by the gravitational merging of two or more spiral galaxies.	
	Irregular Galaxies Irregular galaxies are oddly shaped. Many formed from the gravitational pull of neighboring galaxies. Irregular galaxies contain young stars and have areas of intense star formation.

COLLECT EVIDENCE

How does gravity impact the formation and structure of galaxies?
Record your evidence (C) in the chart at the beginning of the lesson.

Review

Summarize It!

1. **Organize** Build a graphic organizer that shows how a solar system is formed. Make sure to include celestial objects such as stars, planets, moons, asteroids, meteoroids, and comets.

Three-Dimensional Thinking

It was once thought that Earth was the center of the universe. Eventually, it was proven that the planets orbit around the Sun. The illustration shows the path of Earth's orbit around the Sun.

2. Describe the path of Earth if the Sun's gravity were to suddenly stop.

   A Earth would continue to move within its orbit.

   B Earth would move in a straight line towards the Sun.

   C Earth would move in a straight line instead of a curved line.

   D Earth would stop moving and become suspended in one spot.

Halley's Comet orbits the Sun and can be seen from Earth about every 76 years. However, before the work of Sir Isaac Newton and Edmond Halley, comets were thought to pass in a straight line through the solar system. In 1705, Edmond Halley used Newton's laws to determine the gravitational effects of Jupiter and Saturn on a comet that he observed in 1682. Using this information and historical records, he determined that comets seen in 1531 and 1607 were the same comet. Halley correctly calculated the orbit of the comet and predicted its return in 1758.

3. Newton's laws state that all objects exert gravitational force and that objects with more mass exert more force. Which system of objects has the greatest effect on the orbit of Halley's comet?

   A Earth, the Sun, and the Moon

   B the Sun, Jupiter, and Saturn

   C asteroids, meteoroids, and dwarf planets

   D Earth, the Sun, and Saturn

Real-World Connection

4. **Infer** Scientists are actively searching for planets orbiting other stars. They can use gravity to find these planets, called exoplanets. Hypothesize how gravity helps locate distant planets orbiting distant stars.

_____

_____

_____

_____

5. **Infer** Earth has one orbiting moon. Jupiter has 79 confirmed moons. Infer why Earth has fewer moons than Jupiter.

_____

_____

_____

_____

 Still have questions?
Go online to check your understanding about the role of gravity in the universe.

 REVISIT SCIENCE PROBES Do you still agree with the person you chose at the beginning of the lesson? Return to the Science Probe at the beginning of the lesson. Explain why you agree or disagree with that person now.

EXPLAIN THE PHENOMENON Revisit your claim about how gravity impacts components in space. Review the evidence you collected. Explain how your evidence supports your claim.

START PLANNING STEM Module Project Science Challenge

Now that you learned about gravity and its importance in the universe, go to your Module Project to start analyzing the objects in the solar system and the universe and start planning your model.

Objects in Our Solar System

A system consists of parts that make up a whole. What are the different parts that make up our solar system? Put an X next to each of the objects you think are part of our solar system.

_____ planets   _____ the Sun   _____ nearby stars other than the Sun

_____ distant stars   _____ constellations   _____ asteroids

_____ comets   _____ moons   _____ human-made satellites

_____ galaxies   _____ black holes   _____ universe

Explain your thinking. Describe what determines whether an object is part of our solar system.

_____

_____

_____

_____

You will revisit your response to the Science Probe at the end of the lesson.

The Solar System

ENCOUNTER
THE PHENOMENON

How do objects in the solar system compare?

GO ONLINE
Watch the animation *In Our Neighborhood* to to see this phenomenon in action.

The animation shows how scientists organize objects in our solar system using distances. Think of other characteristics that you could use to organize the objects in our solar system. Create a graphic organizer showing your ideas about how objects in the solar system might be compared in the space below.

EXPLAIN
THE PHENOMENON

You examined an animation about the solar system. How do astronomers learn about the objects within the solar system? How do planets, moons, asteroids, and comets compare to each other? Make a claim on how objects in the solar system can be analyzed and interpreted.

CLAIM

The objects of the solar system can be analyzed and interpreted by...

 COLLECT EVIDENCE as you work through the lesson. Then return to these pages to record your evidence.

EVIDENCE

A. What evidence have you discovered for how exploration has helped the advancement of technology?

B. What evidence have you discovered for how the planets in the solar system compare to each other?

MORE EVIDENCE

C. What evidence have you discovered for how moons, asteroids, comets, meteors, and dwarf planets compare to each other?

When you are finished with the lesson, review your evidence. If necessary, based on the evidence, revise your claim.

REVISED CLAIM

The objects of the solar system can be analyzed and interpreted by...

Finally, explain your reasoning for how and why your evidence supports your claim.

REASONING

The evidence I collected supports my claim because...

What objects make up the solar system?

Recall that gas, dust, and stars make up galaxies. Stars are made up of gas that is undergoing fusion at the core. What's in a solar system? A solar system contains at least one star and all the objects that orbit around that star. This includes planets, moons, comets, asteroids, and meteoroids.

Want more information?
Go online to read more about the objects in the solar system and universe.

Go to the Foldables® library to make a Foldable® that will help you take notes while reading this lesson.

How do astronomers observe the solar system?

Ancient observers looking at the night sky saw many stars but only five planets—Mercury, Venus, Mars, Jupiter, and Saturn. The invention of the telescope in the 1600s led to the discovery of additional planets and many other space objects.

Telescopes enable astronomers to observe many more stars than they could with their eyes alone. Telescopes are designed to collect a certain type of electromagnetic wave. Some telescopes detect visible light, and others detect radio waves and microwaves. Astronomers use many kinds of telescopes to study the energy emitted by stars and other objects in space. Let's look at the differences in telescope technology.

INVESTIGATION

Compare the View

Examine the images below.

Jupiter

Saturn

1. These images were taken from a telescope you might use in your backyard. Based on these photos, describe the differences between the two planets.

_____

_____

_____

_____

Now examine the images below.

Jupiter

Saturn

2. These images were taken from a space-based telescope. Based on these photos, describe the differences between the two planets.

_____

_____

_____

_____

Technology Advances In addition to telescopes, space agencies like NASA have sent probes and other exploratory spacecraft into space. For example, the *New Horizons* mission explored the Pluto system within our solar system and will explore objects on the edge of the solar system in the Kuiper belt. Other spacecraft, like the *International Space Station* (*ISS*), house rotating shifts of astronauts for research purposes. The experiments conducted on board the *ISS* will aid future crewed space flights.

The space program requires the development of materials that can withstand the extreme temperatures and pressures of space. Much of this new technology has also been adapted for use in materials on Earth. Some examples of everyday materials that were developed as a result of the space program are satellite television, medical imaging, and home insulation.

COLLECT EVIDENCE

How has space exploration helped the advancement of technology? Record your evidence (A) in the chart at the beginning of the lesson.

How do scientists analyze data about the solar system?

Knowing how scientists obtain data and information on objects, they can then analyze the information and determine the relationships and patterns between characteristics. Let's see how scientists analyze data from the planets in the solar system.

INVESTIGATION

Graphing Characteristics

Scientists collect and analyze data and draw conclusions based on data. Scientists know that some properties of the planets are related. Graphing data makes the relationships easy to identify. Graphing allows different types of data be to seen in relation to one another.

Scientists use different units of measurements for distance within the solar system. One of those units is the astronomical unit (AU). One AU is equal to the average distance between the Sun and Earth, about 149,600,000 km.

Planet	Average Distance from the Sun (AU)	Orbital Period (yr)	Radius of Planet (km)
Mercury	0.39	0.24	2,440
Venus	0.72	0.62	6,051
Earth	1.00	1.0	6,378
Mars	1.52	1.9	3,397
Jupiter	5.20	11.9	71,492
Saturn	9.58	29.4	60,268
Uranus	19.2	84.0	25,559
Neptune	30.1	164.0	24,764

1. You will plot two graphs that explore the relationships in data. The first graph compares a planet's distance from the Sun and its orbital period. The second graph compares a planet's distance from the Sun and its radius. Make a prediction about how these two sets of data are related, if at all. The data are shown in the table.

_____

_____

_____

_____

2. Use the data in the table to plot a line graph showing orbital period versus average distance from the Sun. On the *x*-axis, plot the planet's distance from the Sun. On the *y*-axis, plot the planet's orbital period. Make sure the range of each axis is suitable for the data to be plotted and clearly label each planet's data point.

Orbital Period v. Distance from the Sun

3. Examine the *Orbital Period v. Distance from the Sun* graph. Describe any relationship you observe between a planet's distance from the Sun and its orbital period.

_____

_____

_____

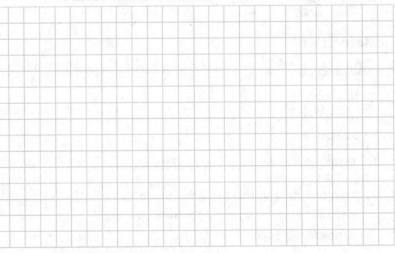

Planet Radius v. Distance from the Sun

4. Use the data in the table to plot a line graph showing planet radius versus average distance from the Sun. On the *x*-axis, plot the planet's distance from the Sun. On the *y*-axis, plot the planet's radius. Make sure the range of each axis is suitable for the data to be plotted and clearly label each planet's data point.

5. Examine the *Planet Radius v. Distance from the Sun* graph. Describe any relationship you observe between a planet's distance from the Sun and its radius. How does this compare to your prediction?

_____

_____

Orbits and Distance from the Sun You've learned that graphing data will sometimes show the relationships between the measured characteristics. The planets' orbital periods are directly related to the average distance from the Sun. You also discovered by graphing data that there is no relationship between the distance from the Sun and the radius of the planet.

How can you model the solar system?

A scale model is a physical representation of something that is much smaller or much larger. Reduced-sized scale models are used to represent and study very large things, such as the solar system. The scale used must reduce the actual size to a size reasonable for the model. Let's investigate how to use a scale model.

 LAB Model the Inner Planets

Safety

Materials

modeling clay
metric ruler

Procedure

1. Read and complete a lab safety form.

2. Use the data in the table for Earth to calculate each model's diameter for the other three planets.

3. Use modeling clay to make a ball that represents the diameter of each planet. Check the diameter with a metric ruler.

Planet	Actual Diameter (km)	Model Diameter (cm)
Mercury	4,879	
Venus	12,103	
Earth	12,756	8.0
Mars	6,792	

Analyze and Conclude

4. **MATH Connection** Explain how you converted actual diameters (km) to model diameters (cm).

_____

_____

5. How do the inner planets' diameters compare?

_____

_____

Now that you have modeled the diameter of the inner planets, let's model the distances in the solar system.

LAB Scale Down

Objects in the solar system are so far apart that astronomers use a larger distance unit. An astronomical unit (AU) is the average distance from Earth to the Sun—about 150 million km. Let's put this scaled measurement into action.

Safety

Materials

register tape   meterstick   masking tape

Procedure

1. Read and complete a lab safety form.

2. First, decide the scale of your solar system. Use the data given in the table on the next page to figure out how far apart the Sun and Neptune would be if a scale of 1 meter = 1 AU is used. Would a solar system based on that scale fit in the space you have available?

_____

3. With your group, determine the scale that results in a model that fits the available space. Larger models are usually more accurate, so choose a scale that produces the largest model that fits in the available space.

Procedure, continued

4. **MATH Connection** Once you have decided on a scale, record the unit you have chosen in the third column of the table. Then fill in the scaled distance for each planet.

Planet	Distance from the Sun (AU)	Distance from the Sun (_____)
Mercury	0.39	
Venus	0.72	
Earth	1.00	
Mars	1.52	
Jupiter	5.20	
Saturn	9.54	
Uranus	19.18	
Neptune	30.06	

5. On register tape, mark the positions of the objects in the solar system based on your chosen scale. Use a length of register tape that is slightly longer than the scaled distance between the Sun and Neptune.

6. Tape the ends of the register tape to a table or the floor. Mark a dot at one end of the paper to represent the Sun. Measure along the tape from the center of the dot to the location of Mercury. Mark a dot at this position and label it *Mercury*. Repeat this process for the remaining planets.

Analyze and Conclude

7. There are many objects in the solar system. These objects have different sizes, structures, and orbits. Examine your scale model of the solar system. How accurate is the model? How could the model be changed to be more accurate?

_____

_____

_____

_____

_____

8. Pluto is a dwarf planet located beyond Neptune. Based on the pattern of distance data for the planets shown in the table, approximately how far from the Sun would you expect to find Pluto? Explain your reasoning.

_____

_____

_____

_____

9. Compare your model with other groups in your class by taping them all side-by-side. Discuss any major differences in your models. Discuss the difficulties in making the scale models much smaller.

_____

_____

_____

_____

_____

_____

_____

THREE-DIMENSIONAL THINKING

How can you build a scale model of the solar system that accurately shows both planetary diameters and distances and the locations of the asteroid belt and dwarf planets? Describe how you would create this model.

_____

_____

_____

_____

Digging Deeper

You have created a model of the planetary distances in our solar system. As a class, you will determine the scale diameter of each of the planets and then research and create a model of their inner layers and the composition of those layers.

Work together and choose a planet to model. After you've chosen a planet, research that planet and its structure and composition. Take notes about your planet below.

1. What is your planet composed of?

_____

_____

2. How is your planet structured?

_____

_____

_____

3. What surface features are present on your planet?

_____

_____

_____

Using materials of your choice, begin constructing your model. You may also create a model on a computer program that can create a 3-D model.

Once you have created your model, share your model with the class. Make sure you are able to answer questions about the surface features and composition of your planet.

COLLECT EVIDENCE

How do the planets in the solar system compare to each other? Record your evidence (B) in the chart at the beginning of the lesson.

Read a Scientific Text

Astronomers learn more about the terrestrial planets using space probes.

CLOSE READING

Inspect

Read the passage *The Incredible Shrinking Mercury is Active After All.*

Find Evidence

Reread the fourth paragraph. Underline the surface features that are present on Mercury.

Make Connections

Communicate With your partner, research surface features on Venus and Mars and compare those with the surface features on Earth and Mercury.

PRIMARY SOURCE

The Incredible Shrinking Mercury is Active After All

It's small, it's hot, and it's shrinking. New NASA-funded research suggests that Mercury is contracting even today, joining Earth as a tectonically active planet.

Images obtained by NASA's MErcury Surface, Space ENvironment, GEochemistry, and Ranging (MESSENGER) spacecraft reveal previously undetected small fault scarps—cliff-like landforms that resemble stair steps. These scarps are small enough that scientists believe they must be geologically young, which means Mercury is still contracting and that Earth is not the only tectonically active planet in our solar system, as previously thought. ...

"The young age of the small scarps means that Mercury joins Earth as a tectonically active planet, with new faults likely forming today as Mercury's interior continues to cool and the planet contracts," said lead author Tom Watters, Smithsonian senior scientist at the National Air and Space Museum in Washington, D.C.

Large fault scarps on Mercury were first discovered in the flybys of Mariner 10 in the mid-1970s and confirmed by MESSENGER, which found the planet closest to the sun was shrinking. The large scarps were formed as Mercury's interior cooled, causing the planet to contract and the crust to break and thrust upward along faults making cliffs up to hundreds of miles long and some more than a mile (over one-and-a-half kilometers) high.

In the last 18 months of the MESSENGER mission, the spacecraft's altitude was lowered, which allowed the surface of Mercury to be seen at much higher resolution. These low-altitude images revealed small fault scarps that are orders of magnitude smaller than the larger scarps. The small scarps had to be very young, investigators say, to survive the steady bombardment of meteoroids and comets. They are comparable in scale to small, young lunar scarps that are evidence Earth's moon is also shrinking.

This active faulting is consistent with the recent finding that Mercury's global magnetic field has existed for billions of years and with the slow cooling of Mercury's still hot outer core. It's likely that the smallest of the terrestrial planets also experiences Mercury-quakes—something that may one day be confirmed by seismometers. ...

Source: National Aeronautics and Space Administration

What are the other objects in the solar system?

Different types of objects orbit the Sun. These objects include dwarf planets, asteroids, meteors, and comets. Unlike the Sun, these objects don't emit light but only reflect the Sun's light. What are the features, compositions, and locations of these objects? Let's investigate!

Moons of the Outer Planets

Jupiter has 79 confirmed moons. Saturn has 53 named moons and 8 provisional moons. Uranus has 27 moons while Neptune only has 13 named moons with one provisional moon. Why do the outer planets have so many moons compared to the inner planets?

The moons of the outer planets range in diameter from 2 km to 5,268 km. The largest moon in the solar system is Jupiter's Ganymede, which is larger than the planet Mercury. How can a moon be larger than a planet? Remember, to be classified as a moon, an object has to orbit a planet, and not the Sun.

Jupiter

Saturn

Moons of the Outer Planets			
Planet	Number of Moons	Largest Moons	Moon Sizes
Jupiter	79 confirmed	Ganymede, Callisto, Io, Europa	These four moons are planet-sized. Ganymede is the solar system's largest moon, with a diameter of 5,268 km.
Saturn	at least 53	Titan, Rhea, Iapetus, Dione	Titan is planet-sized at 5,150 km in diameter.
Uranus	at least 27	Titania, Oberon, Umbriel, Ariel	Titania is 1,578 km in diameter, while tiny Cordelia is 26 km in diameter.
Neptune	at least 13	Triton, Proteus, Nereid, Larissa	Triton is 2,700 km in diameter.

1. Select the largest moon of each outer planet. Then, draw the moons in order from largest to smallest. Identify each moon and its planet. Create a scale so you can keep your drawings proportional. Be sure to show your scale.

2. With a partner, research each of the identified moons. Compare and contrast a chosen characteristic—surface features, composition, structure, distance from the planet—for each of the moons. Use your Science Notebook if you need more room.

Moons Many moons of the outer planets are small with irregular shapes and unusual orbits. Some scientists think that these are captured asteroids, meaning the gravity of the planet pulled the object into its orbit. Captured asteroids are natural satellites, but they did not form by accretion as regular spherical moons probably did.

Dwarf Planets Scientists classify some objects in the solar system as dwarf planets. A **dwarf plane**t is a spherical object that orbits the Sun. It is not a moon of another planet and is in a region of the solar system where there are many objects orbiting near it. But, unlike a planet, a dwarf planet does not have more mass than objects in nearby orbits. Dwarf planets include Ceres (SIHR eez), Eris (IHR is), Pluto, and Makemake (MAH kay MAH kay). Dwarf planets are made of rock and ice and are much smaller than Earth.

Ceres

Pluto

Other Solar System Objects	
Asteroids **Asteroids** are chunks of rock and ice that never clumped together like the rocks and ice that formed the inner planets. Some astronomers suggest that Jupiter's strong gravity might have caused the chunks to collide so violently that they broke apart instead of sticking together. This means that asteroids are objects left over from the formation of the solar system. Most asteroids are found in the asteroid belt between Mars and Jupiter.	
Comets **Comets** are mixtures of rock, ice, and dust. The particles in a comet are loosely held together by the gravitational attractions among the particles. Comets orbit the Sun in long elliptical orbits.	
Meteoroids, Meteors, Meteorites **Meteoroids** are small, rocky particles that move through space. Most meteoroids are only about as big as a grain of sand. When meteoroids pass through Earth's atmosphere, they burn up because of friction and are seen as streaks of light in the sky. This is called a **meteor.** If a meteor doesn't completely burn up, it can impact the planet's surface. At that point, it is called a **meteorite.**	

COLLECT EVIDENCE

How do the moons, asteroids, comets, meteors, and dwarf planets compare to each other? Record your evidence (C) in the chart at the beginning of the lesson.

🅥 **GO ONLINE** for additional opportunities to explore!

Investigate more about dwarf planets and other objects in the solar system by completing one or both of these activities.

☐ **Read** the **Scientific Text** *Dwarf Planets and Other Objects— Comets.*

OR

☐ **Complete** the **Lab** *How might asteroids and moons form?*

Copyright © McGraw-Hill Education (t)NASA/JPL–Caltech/UCAL/MPS/ DLR/IDA, (c)J. Linder/ESO, (b)Luis Argerich/Science Source

Pluto

What in the world is it?

Since Pluto's discovery in 1930, students have learned that the solar system has nine planets. But in 2006, the number of planets was changed to eight. What happened?

Astrophysicists at the American Museum of Natural History in New York City were among those that first questioned Pluto's classification as a planet. Pluto orbits the Sun at the outer edge of the solar system in a region called the Kuiper Belt. It shares this region with thousands of other objects. One of the largest objects, known as Eris, is almost the same size as Pluto. Shouldn't Eris be classified as a planet too?

This question spurred the International Astronomical Union (IAU) to set up a committee to define what constitutes a planet. In 2006, the IAU came up with three rules for being considered a planet in our solar system. First, the object has to be massive enough for gravity to shape it into a sphere. Secondly, it has to be in orbit around the Sun. Finally, the object must have a clear orbit devoid of other objects. In other words, it has to have its own "zone." Because Pluto has other objects in its zone, it was demoted from its status as a planet and reclassified as a dwarf planet. And while Pluto may have lost its rank as the smallest planet, it still is "King of the Kuiper Belt."

This illustration shows what Pluto might look like if you were standing on one of its moons.

Pluto TIME LINE

1930
Astronomer Clyde Tombaugh discovers a ninth planet, Pluto.

1992
The first Kuiper Belt object is discovered.

2005
Eris—an object the size of Pluto—is discovered in the Kuiper Belt.

January 2006
NASA launched the *New Horizons* spacecraft that will study Pluto, its moon, and other Kuiper Belt objects.

August 2006
Pluto is reclassified as a dwarf planet.

July 2015
New Horizons reached Pluto and began gathering data.

It's Your Turn

MAKE A LIST What did scientists learn about Pluto when the *New Horizons* spacecraft visited the Pluto system in 2015? Make a list with three things they discovered.

Review

Summarize It!

1. **Construct** a graphic organizer that compares and contrasts objects in the solar system.

Three-Dimensional Thinking

● NASA has been sending exploration missions to Mars for almost 50 years. The length of time it has taken the different spacecraft to reach Mars ranges from 150 days to 360 days. In order to make it possible for humans to travel to Mars, the travel time must be reduced.

2. Assume the different spacecraft were using the same amount of fuel at the same rate. Why are there such long differences in travel time?

   A The distance of Mars from the Sun changes.

   B The distance from Earth to Mars changes.

   C The position of Earth from the Sun changes.

   D The shape of the planet's orbit changes.

The chart below shows the surface temperatures of the outer planets.

Planet	Surface Temperature
Neptune	−214°C
Uranus	−216°C
Jupiter	−148°C
Saturn	−178°C

3. What can you conclude about the outer planets based on the data table?

   A If Neptune were closer to the Sun, it would be cooler.

   B Saturn is closer to the Sun than any of the other outer planets.

   C The coldest planet is Uranus.

   D The surface temperature of Earth is between that of Neptune and Jupiter.

Real-World Connection

4. Evaluate Think about the ways in which scientists observe, study, and explore space. What technology, other than the examples given to you in this lesson, has been added to everyday life because of space exploration?

_____

_____

_____

_____

_____

_____

_____

_____

_____

_____

_____

Copyright © McGraw-Hill Education   Purestock/SuperStock

 Still have questions?
Go online to check your understanding about objects in our solar system.

 REVISIT PAGE KEELEY **SCIENCE PROBES**
Do you still agree with the answers you chose at the beginning of the lesson? Return to the Science Probe at the beginning of the lesson. Explain why you agree or disagree with those answers now.

EXPLAIN THE PHENOMENON

Revisit your claim about how objects in our solar system can be analyzed and interpreted. Review the evidence you collected. Explain how your evidence supports your claim.

PLAN AND PRESENT
STEM Module Project Science Challenge

Now that you learned about the objects in the solar system, go to your Module Project to continue analyzing the objects in the solar system and the universe, finish building your model, and give your presentation.

Wanted: Space Investigator

You receive an email with the subject line "Wanted: Space Investigators." The email reads:

"The National Aeronautics and Space Administration (NASA) has selected you to manage a team of space investigators to analyze and model objects in space. Your team will deliver a presentation to government officials who want to know more about our galaxy and solar system, and the technology used to explore space."

In 3 ... 2 ... 1 ... begin your mission, Space Investigator!

Planning After Lesson 1

How can you indicate the relative spatial scales of our solar system and galaxy in a model?

Gather the data you need for your model and organize it in your Science Notebook. Cite the sources of your data.

How will you incorporate the role of gravity in your model?

Planning After Lesson 2

Review the data you gathered in Lesson 2 about objects in our solar system. What additional data do you need for your model and analysis of solar system objects?

Gather your data and organize it in the space below. Be sure to cite the sources of your data. Use your Science Notebook if you need more space.

Develop Your Model

Review the planning you did after Lessons 1 and 2, then develop a model that describes the role of gravity in the solar system and Milky Way galaxy. You will also need to analyze and interpret your data in order to incorporate elements in your model that show the scale properties of objects in our solar system. Sketch your model below. List the materials you will use.

Evaluate Your Model

Model Elements	Descriptions
Components (What are the different parts of my model?)	
Relationships (How do the components of my model interact?)	
Connections (How does my model help me understand the phenomenon?)	

Present Your Model

After you have observed and evaluated your model, answer the questions below.

Based on your data, identify one advance in solar system science made possible by improved engineering. Conversely, identify one advance in engineering made possible by advances in solar system science.

Use your model to identify patterns in the characteristics of solar system objects. Describe one pattern below.

Use your model to identify patterns or draw conclusions about the scale of solar system objects. Write one pattern or conclusion below.

Congratulations! You've completed the Science Challenge requirements.

Module Wrap-Up

REVISIT
THE PHENOMENON

Using the model and analysis of the objects in the solar system, explain the relationship between solar system objects and other objects in the universe.

OPEN INQUIRY

If you had to ask one question about what you studied, what would it be?

_____

_____

_____

_____

Plan and conduct an investigation to answer this question.

Glossary

Pronunciation Key

Use the following key to help you sound out words in the glossary.

a	back (BAK)	Ew	food (FEWD)
ay	day (DAY)	yoo	pure (PYOOR)
ah	father (FAH thur)	yew	few (FYEW)
ow	flower (FLOW ur)	uh	comma (CAH muh)
ar	car (CAR)	u (+ con)	rub (RUB)
E	less (LES)	sh	shelf (SHELF)
ee	leaf (LEEF)	ch	nature (NAY chur)
ih	trip (TRIHP)	g	gift (GIHFT)
i (i + com + e)	idea (i DEE uh)	J	gem (JEM)
oh	go (GOH)	ing	sing (SING)
aw	soft (SAWFT)	zh	vision (VIH zhun)
or	orbit (OR buht)	k	cake (KAYK)
oy	coin (COYN)	s	seed, cent (SEED)
oo	foot (FOOT)	z	zone, raise (ZOHN)

English ——— A ——— **Español**

asteroid/dwarf planet   **asteroide/planeta enano**

asteroid: a small, rocky object that orbits the Sun.

asteroide: objeto pequeño y rocoso que orbita el Sol.

C

carrying capacity: the largest number of individuals of one species that an ecosystem can support over time.

comet: a small, rocky, icy object that orbits the Sun.

capacidad de carga: número mayor de individuos de una especie que un medioambiente puede mantener.

cometa: objeto pequeño, rocoso y helado que orbita el Sol.

D

deforestation: the removal of large areas of forests for human purposes.

dwarf planet: an object that orbits the Sun and is nearly spherical in shape, but shares its orbital path with other objects of similar size.

deforestación: eliminación de grandes áreas de bosques con propósitos humanos.

planeta enano: objeto de forma casi esférica que orbita el Sol y que comparte el recorrido de la órbita con otros objetos de tamaño similar.

E

ecological footprint: the area of Earth's productive land and water required to supply the resources that an individual demands as well as to absorb the wastes that the individual produces.

equinox: when Earth's rotation axis is tilted neither toward nor away from the Sun.

huella ecológica: área productiva de la tierra y el agua de la Tierra requerida para proveer los recursos que un individuo demanda y también para absorber los desechos que el individuo produce.

equinoccio: cuando el eje de rotación de la Tierra se inclina sin acercarse ni alejarse del Sol.

G

galaxy: a huge collection of stars, gas, and dust.

gravity: an attractive force that exists between all objects that have mass.

galaxia: conjunto enorme de estrellas, gas, y polvo.

gravedad: fuerza de atracción que existe entre todos los objetos que tienen masa.

L

lunar eclipse: an occurrence during which the Moon moves into Earth's shadow.

eclipse lunar: ocurrencia durante la cual la Luna se mueve hacia la zona de sombra de la Tierra.

M

meteor: a meteoroid that has entered Earth's atmosphere and produces a streak of light.

meteorite: a meteoroid that strikes a planet or a moon.

meteoroid: a small, rocky particle that moves through space.

meteoro: meteorito que ha entrado a la atmósfera de la Tierra y produce un haz de luz.

meteorito: meteoroide que impacta un planeta o una luna.

meteoroide: partícula rocosa pequeña que se mueve por el espacio.

N

natural resource: part of the environment that supplies material useful or necessary for the survival of living things.

recurso natural: parte del medio ambiente terrestre que proporcionan materiales útiles o necesarios para la supervivencia de los organismos vivos.

O

orbit: the path an object follows as it moves around another object.

órbita: trayectoria que un objeto sigue a medida que se mueve alrededor de otro objeto.

P

penumbra: the lighter part of a shadow where light is partially blocked.

phase: the lit part of the Moon or a planet that can be seen from Earth.

population: all the organisms of the same species that live in the same area at the same time.

penumbra: parte más clara de una sombra donde la luz se bloquea parcialmente.

fase: parte iluminada de la Luna o de un planeta que se ve desde la Tierra.

población: todos los organismos de la misma especie que viven en la misma área al mismo tiempo.

R

reclamation: a process in which mined land is restored with soil and replanted with vegetation.

revolution: the orbit of one object around another object.

rotation: the spin of an object around its axis.

rotation axis: the line on which an object rotates.

recuperación: proceso por el cual las tierras explotadas se deben recubrir con suelo y se deben replantar con vegetación.

revolución: movimiento de un objeto alrededor de otro objeto.

rotación: movimiento giratorio de un objeto sobre su eje.

eje de rotación: línea sobre la cual un objeto rota.

Glossary Glosario

S

solar eclipse: an occurrence during which the Moon's shadow appears on Earth's surface.

solstice: when Earth's rotation axis is tilted directly toward or away from the Sun.

eclipse solar: acontecimiento durante el cual la sombra de la Luna aparece sobre la superficie de la Tierra.

solsticio: cuando el eje de rotación de la Tierra se inclina acercándose o alejándose del Sol.

U

umbra: the central, darker part of a shadow where light is totally blocked.

urbanization: the development of towns and cities on previously natural areas.

umbra: parte central más oscura de una sombra donde la luz está completamente bloqueada.

urbanización: desarollo de los pueblos y las ciudades en zonas previamente naturales

W

waning phases: phases of the Moon during which less of the Moon's near side is lit each night.

waxing phases: phases of the Moon during which more of the Moon's near side is lit each night.

fases menguantes: fases de la Luna durante las cuales el lado cercano de la Luna está menos iluminado cada noche.

fases crecientes: fases de la Luna durante las cuales el lado cercano de la Luna está más iluminado cada noche.